Stochastic Processes
with R

CHAPMAN & HALL/CRC
Texts in Statistical Science Series

Joseph K. Blitzstein, *Harvard University, USA*
Julian J. Faraway, *University of Bath, UK*
Martin Tanner, *Northwestern University, USA*
Jim Zidek, *University of British Columbia, Canada*

Recently Published Titles

Bayesian Thinking in Biostatistics
Gary L. Rosner, Purushottam W. Laud, and Wesley O. Johnson

Linear Models with Python
Julian J. Faraway

Modern Data Science with R, Second Edition
Benjamin S. Baumer, Daniel T. Kaplan, and Nicholas J. Horton

Probability and Statistical Inference
From Basic Principles to Advanced Models
Miltiadis Mavrakakis and Jeremy Penzer

Bayesian Networks
With Examples in R, Second Edition
Marco Scutari and Jean-Baptiste Denis

Time Series
Modeling, Computation, and Inference, Second Edition
Raquel Prado, Marco A. R. Ferreira and Mike West

Foundations of Statistics for Data Scientists
With R and Python
Alan Agresti and Maria Kateri

Fundamentals of Causal Inference
With R
Babette A. Brumback

Stochastic Processes with R
An Introduction
Olga Korosteleva

For more information about this series, please visit: https://www.routledge.com/Chapman--HallCRC-Texts-in-Statistical-Science/book-series/CHTEXSTASCI

Stochastic Processes with R

An Introduction

Olga Korosteleva

CRC Press
Taylor & Francis Group
Boca Raton London New York

CRC Press is an imprint of the
Taylor & Francis Group, an **informa** business
A CHAPMAN & HALL BOOK

First edition published 2022
by CRC Press
6000 Broken Sound Parkway NW, Suite 300, Boca Raton, FL 33487-2742

and by CRC Press
2 Park Square, Milton Park, Abingdon, Oxon, OX14 4RN

CRC Press is an imprint of Taylor & Francis Group, LLC

ISBN: 978-1-032-15373-5 (hbk)
ISBN: 978-1-032-15473-2 (pbk)
ISBN: 978-1-003-24428-8 (ebk)

DOI: 10.1201/9781003244288

Publisher's note: This book has been prepared from camera-ready copy provided by the author.

Access the Solutions Manual for instructors: https://www.routledge.com/9781032153735

Contents

Preface

This book was written as a textbook for an undergraduate, senior-level course on random processes for Statistics majors. The presentation is meant to be light yet sufficiently mathematical, with good, interesting, scientific applications. More advanced topics, such as sigma algebras, martingales, general renewal processes, Levy processes, and stochastic calculus, are deliberately avoided. The knowledge of statistics is limited to a linear regression, goodness of fit test, and point estimation.

There are nine chapters in this book, covering Markov chain, random walk, Poisson processes (homogeneous, nonhomogeneous, compound, and conditional), birth-and-death process, branching process, and Brownian motion. Each chapter gives just enough theory in the form of definitions, propositions, remarks, and examples, followed by a section on simulation of trajectories. Finally, applications of processes are presented. At the end of each chapter, a collection of exercises is included. Some of the exercises are theoretical whereas some others require calculation and simulation.

The R software is used throughout. Complete codes and relevant outputs are given in the text. The website that accompanies this book

https://home.csulb.edu/∼okoroste/stochprocesses.html

contains complete R codes for all examples and applications, and the data sets in .csv format that are used in applications and/or required for certain exercises. A complete solutions manual is also available to instructors upon request at https://www.routledge.com/9781032153735.

Last but not least, I would like to thank John Peach who helped me to simulate a trajectory of a branching process by means of self-referencing functions.

Respectfully,
The author

Author

Olga Korosteleva, PhD, is a professor of statistics in the Department of Mathematics and Statistics at California State University, Long Beach (CSULB). She earned her Bachelor's degree in mathematics in 1996 from Wayne State University in Detroit, and her PhD in statistics from Purdue University in West Lafayette, Indiana, in 2002. Since then she has been teaching statistics and mathematics courses at CSULB.

1

Stochastic Process, Discrete-time Markov Chain

1.1 Definition of Stochastic Process

A *stochastic process* $\{X(t),\ t \in T\}$ is a collection of random variables indexed by a parameter t that belongs to a set T. The parameter t is often referred to as *time*, and the value $X(t)$ is the *state* of the process at time t. The set T is called the *index set* of the process.

The *state space* S of a stochastic process is the set of all possible values of $X(t)$, for any $t \in T$.

If T is a countable set, the stochastic process is termed a *discrete-time process* (or, simply, a *discrete process*). Otherwise, it is called a *continuous-time process*.

1.2 Discrete-time Markov Chain

A *discrete-time Markov chain*[1] is a discrete-time stochastic process $\{X_n,\ n = 0, 1, 2, etc.\}$ which state space S is finite or countably infinite and such that

$$\mathbb{P}(X_{n+1} = j \mid X_0 = i_0,\ X_1 = i_1,\ \ldots,\ X_n = i) = \mathbb{P}(X_{n+1} = j \mid X_n = i) = P_{ij}, \tag{1.1}$$

that is, the conditional probability of the process being in state j at time $n+1$ given all the previous states depends only on the last-known position (state

---

[1] Introduced in Markov, A. A. (1913). "An example of statistical investigation of the text Eugene Onegin concerning the connection of samples in chains." (In Russian.) *Bulletin of the Imperial Academy of Sciences of St. Petersburg*, 7(3): 153 − 162.

DOI: 10.1201/9781003244288-1

1

i at time n). This property is called the *Markovian property* (or the *Markov property*). The probability P_{ij} is called the *one-step transition probability* of the Markov chain. Note that it is constant for given states i and j.

In Markov chains, one-step transition probabilities are typically aggregated into a *one-step transition probability matrix*

$$\mathbf{P} = \begin{bmatrix} P_{00} & P_{01} & P_{02} & \cdots \\ P_{10} & P_{11} & P_{12} & \cdots \\ & \cdots & & \\ P_{i0} & P_{i1} & P_{i2} & \cdots \\ & \cdots & & \end{bmatrix}.$$

EXAMPLE 1.1. Consider a Markov chain with the state space $S = \{1, 2, 3\}$ and transition probability matrix

$$\mathbf{P} = \begin{matrix} & \begin{matrix} 1 & 2 & 3 \end{matrix} \\ \begin{matrix} 1 \\ 2 \\ 3 \end{matrix} & \begin{bmatrix} 0.7 & 0.1 & 0.2 \\ 0.0 & 0.6 & 0.4 \\ 0.5 & 0.2 & 0.3 \end{bmatrix} \end{matrix}.$$

(a) Starting in state 1, the Markov chain returns to it in one step with probability 0.7, or transitions to state 2 with probability 0.1, or transitions to state 3 with probability 0.2. From state 2, the chain cannot transition to state 1 in a single step since the probability of this event is 0. It can, however, return to state 2 or move to state 3 with probabilities 0.6 and 0.4, respectively. If the chain is in state 3, it will transition to states 1 or 2 with respective probabilities 0.5 and 0.2 or will loop back to state 3 with a probability of 0.3. Note that since the chain must transition to some state, the probabilities in each row necessarily sum up to 1.

(b) Conditional probabilities can be computed using the Markov property. For example, $\mathbb{P}(X_3 = 1 \mid X_0 = 1, X_1 = 2, X_2 = 3) = \mathbb{P}(X_3 = 1 \mid X_2 = 3) = P_{31} = 0.5$.

(c) Joint probabilities can be computed by conditioning and using the Markov property. For example, we want to compute the probability that a Markov chain starts in state 1 at time 0, then transitions into state 2, and then into state 3. We obtain $\mathbb{P}(X_0 = 1, X_1 = 2, X_2 = 3) = \mathbb{P}(X_2 = 3 \mid X_0 = 1, X_1 = 2) \, \mathbb{P}(X_0 = 1, X_1 = 2) = \mathbb{P}(X_2 = 3 \mid X_1 = 2) \, \mathbb{P}(X_1 = 2 \mid X_0 = 1) \mathbb{P}(X_0 = 1) = P_{23} \cdot P_{12} \cdot \mathbb{P}(X_0 = 1) = (0.4)(0.1)(1) = 0.04.$ □

1.3   Chapman-Kolmogorov Equations

Consider a Markov chain with finite or countably infinite state space $S = \{0, 1, 2, \dots\}$. For fixed states i and j, the *n-step transition probability* P_{ij}^n is the probability that a process that is currently in state i will be in state j after n transitions, that is, for any time $m \geq 0$, $P_{ij}^n = \mathbb{P}(X_{m+n} = j | X_m = i)$.

Denote the *n-step transition probability matrix* by $\mathbf{P}^{(n)}$. As a special case with $n = 1$, $\mathbf{P}^{(1)} = \mathbf{P}$.

Next, we will show that $\mathbf{P}^{(n)} = \mathbf{P}^n$, which indicates that finding the n-step transition probability matrix is tantamount to multiplying the one-step transition probability matrix by itself n times.

The proof is based on the *Chapman-Kolmogorov equations* which assert that for all positive integers n and m, $\mathbf{P}^{(n+m)} = \mathbf{P}^{(n)} \cdot \mathbf{P}^{(m)}$, or, equivalently, $P_{ij}^{n+m} = \sum_{k=0}^{\infty} P_{ik}^n P_{kj}^m$, for any states i and j. Indeed, by the definition of conditional probability,

$$P_{ij}^{n+m} = \mathbb{P}(X_{n+m} = j | X_0 = i) = \frac{\mathbb{P}(X_{n+m} = j, X_0 = i)}{\mathbb{P}(X_0 = i)}.$$

Next, we fix state k in which trajectory of the Markov chain is located after n transitions and sum up the probabilities with respect to k, obtaining:

$$P_{ij}^{n+m} = \sum_{k=0}^{\infty} \frac{\mathbb{P}(X_{n+m} = j, X_n = k, X_0 = i)}{\mathbb{P}(X_0 = i)},$$

which by the definition of conditional probability is equal to

$$= \sum_{k=0}^{\infty} \frac{\mathbb{P}(X_{n+m} = j | X_n = k, X_0 = i)\mathbb{P}(X_n = k, X_0 = i)}{\mathbb{P}(X_0 = i)},$$

and, applying the definition of conditional probability again, we get

$$= \sum_{k=0}^{\infty} \mathbb{P}(X_{n+m} = j | X_n = k, X_0 = i)\mathbb{P}(X_n = k | X_0 = i).$$

Finally, applying the Markov property, we can omit from the history all but the latest known state that we condition on, and deduce that

$$P_{ij}^{n+m} = \sum_{k=0}^{\infty} \mathbb{P}(X_{n+m} = j | X_n = k)\mathbb{P}(X_n = k | X_0 = i) = \sum_{k=0}^{\infty} P_{kj}^m P_{ik}^n.$$

This completes the proof. Now, as a corollary of the Chapman-Kolmogorov equations, we will show that $\mathbf{P}^{(n)} = \mathbf{P}^n$. Applying the method of mathematical induction, we first check that the statement is true for small values of n: $\mathbf{P}^{(2)} = \mathbf{P}^{(1+1)} = \mathbf{P}^{(1)} \cdot \mathbf{P}^{(1)} = \mathbf{P} \cdot \mathbf{P} = \mathbf{P}^2$, and $\mathbf{P}^{(3)} = \mathbf{P}^{(2)} \cdot \mathbf{P} = \mathbf{P}^3$. Assuming further that the statement holds for $n-1$, we prove it for n. We write, $\mathbf{P}^{(n)} = \mathbf{P}^{(n-1)} \cdot \mathbf{P} = \mathbf{P}^{n-1} \cdot \mathbf{P} = \mathbf{P}^n$.

EXAMPLE 1.2. In Example 1.1, we considered a Markov chain with the one-step transition probability matrix

$$\mathbf{P} = \begin{bmatrix} 0.7 & 0.1 & 0.2 \\ 0.0 & 0.6 & 0.4 \\ 0.5 & 0.2 & 0.3 \end{bmatrix}.$$

(a) Suppose we would like to find a three-step transition probability matrix. We can compute

$$\mathbf{P}^{(3)} = \mathbf{P}^3 = \begin{bmatrix} 0.7 & 0.1 & 0.2 \\ 0.0 & 0.6 & 0.4 \\ 0.5 & 0.2 & 0.3 \end{bmatrix} \cdot \begin{bmatrix} 0.7 & 0.1 & 0.2 \\ 0.0 & 0.6 & 0.4 \\ 0.5 & 0.2 & 0.3 \end{bmatrix} \cdot \begin{bmatrix} 0.7 & 0.1 & 0.2 \\ 0.0 & 0.6 & 0.4 \\ 0.5 & 0.2 & 0.3 \end{bmatrix}$$

$$= \begin{bmatrix} 0.59 & 0.17 & 0.24 \\ 0.20 & 0.44 & 0.36 \\ 0.50 & 0.23 & 0.27 \end{bmatrix} \cdot \begin{bmatrix} 0.7 & 0.1 & 0.2 \\ 0.0 & 0.6 & 0.4 \\ 0.5 & 0.2 & 0.3 \end{bmatrix} = \begin{bmatrix} 0.533 & 0.209 & 0.258 \\ 0.320 & 0.356 & 0.324 \\ 0.485 & 0.242 & 0.273 \end{bmatrix}.$$

(b) Now we can compute probabilities that require the knowledge of the three-step transition probability matrix. For instance, if $\mathbb{P}(X_0 = 1) = 1$, we can calculate $\mathbb{P}(X_0 = 1, X_1 = 2, X_4 = 3) = \mathbb{P}(X_4 = 3 \mid X_0 = 1, X_1 = 2) \, \mathbb{P}(X_1 = 2 \mid X_0 = 1) \mathbb{P}(X_0 = 1) = \mathbb{P}(X_4 = 3 \mid X_1 = 2) \, \mathbb{P}(X_1 = 2 \mid X_0 = 1) \mathbb{P}(X_0 = 1) = P_{23}^{(3)} \cdot P_{12} \cdot \mathbb{P}(X_0 = 1) = (0.324)(0.1)(1) = 0.0324.$ □

Further, suppose we are given the distribution of the initial state $p_i^0 = \mathbb{P}(X_0 = i)$. Conditioning on this distribution, we can obtain the *unconditional* distribution of the state at time n as

$$p_j^n = \mathbb{P}(X_n = j) = \sum_{i=0}^{\infty} \mathbb{P}(X_n = j \mid X_0 = i) \mathbb{P}(X_0 = i) = \sum_{i=0}^{\infty} P_{ij}^n \, p_i^0.$$

In the matrix form, this equation becomes

$$(p_1^n, p_2^n, \dots) = (p_1^0, p_2^0, \dots) \, \mathbf{P}^{(n)}.$$

EXAMPLE 1.3. Suppose that in Example 1.1 the initial states are equiprobable, that is, $p_1 = p_2 = p_3 = 1/3$. Then, after three transitions, the probability that the chain ends up in state 1 is $\mathbb{P}(X_3 = 1) = \sum_{i=1}^{3} P_{i1}^{(3)} p_i = (0.533)(1/3) + (0.320)(1/3) + (0.485)(1/3) = 0.446$. Note that in these calculations, the three-step transition probabilities come from the first column of the three-step transition probability matrix obtained in Example 1.2. Likewise, the probability that after three steps the chain will be in state 2 can be found as $\mathbb{P}(X_3 = 2) = \sum_{i=1}^{3} P_{i2}^{(3)} p_i = (0.209)(1/3) + (0.356)(1/3) + (0.242)(1/3) = 0.269$. The unconditional probability of state 3 may be obtained by similar calculations $\mathbb{P}(X_3 = 3) = (0.258)(1/3) + (0.324)(1/3) + (0.273)(1/3) = 0.285$, or by subtraction from 1, $\mathbb{P}(X_3 = 3) = 1 - 0.4646 - 0.269 = 0.285$.

These calculations may be written in the matrix form as follows:

$$(1/3, 1/3, 1/3) \begin{bmatrix} 0.533 & 0.209 & 0.258 \\ 0.320 & 0.356 & 0.324 \\ 0.485 & 0.242 & 0.273 \end{bmatrix} = (0.446, 0.269, 0.285). \quad \square$$

1.4 Classification of States

Consider a discrete-time Markov chain $\{X_n, n \geq 0\}$. We say that starting in state i, the chain *ever enters state j* if for some $n \geq 0$, $X_n = j$, that is,

$$\{\text{chain ever enters state } j \mid \text{it starts in state } i\} = \bigcup_{n=0}^{\infty} \{X_n = j \mid X_0 = i\}.$$

A state j is called *accessible* from state i if, starting in i, the chain will ever enter state j with a positive probability. To denote that state j is accessible from state i, we write $i \to j$.

Further, if two states are accessible from each other, we say that they *communicate*, and denote it as $i \leftrightarrow j$. Communication property is an equivalence relation. Indeed, (i) it is reflexive since any state communicates with itself in 0 steps, (ii) it is symmetric by definition: if state i communicates with state j, then state j communicates with state i, and (iii) it is transitive because if state i communicates with state j, and state j communicates with state k, then state i communicates with state k.

Now, since communication is an equivalence relation, it means that communication is a *class property*: all states that communicate with each other belong to the same class, and the state space of a Markov chain may be partitioned into classes. If there is only one class, the chain is termed *irreducible*.

Next, a state is called *recurrent* if with probability 1 the chain ever reenters that state. Otherwise, the state is called *transient*. Any state is either recurrent or transient. Furthermore, at least one state of a finite-state Markov chain must be recurrent because if all states are transient, after the chain leaves all the states it has no state to go to. Similar to the communication relation, recurrence and transience are class properties, and the entire class is called *recurrent* or *transient*.

In addition, a state is called *absorbing* if the chain cannot leave it once it enters it. An *absorbing Markov chain* has at least one absorbing state. A state is termed *reflecting* if once the chain leaves it, it cannot return to it.

Finally, the *period* d of a state i is the number such that, starting in i, the chain can return to i only in the number of steps that are multiples of d. A state with period $d = 1$ is called *aperiodic*. Periodicity is a class property.

For a reflecting state, the period is infinite, since the chain never comes back to this state. Absorbing states necessarily have loops and thus are aperiodic states.

EXAMPLE 1.4. Referring back to Example 1.1, we first draw a diagram of the Markov chain. The R code that produces the diagram will be displayed in Example 1.8.

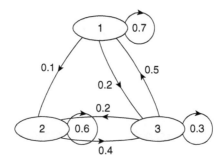

As seen in the diagram, state 2 can be reached from state 1 directly, and states 2 and 3 are directly accessible from each other. Also, state 1 is accessible from state 2 through state 3. Therefore, all three states communicate, and thus the chain has a single class and is irreducible. Since there is only one class, it must be recurrent. Also, the chain has loops for every state, thus it can return to every state in one step and is aperiodic. □

EXAMPLE 1.5. Consider a Markov chain with the state space $\{1, 2, 3, 4, 5, 6\}$ and the one-step transition probability matrix

	1	2	3	4	5	6
1	0.3	0.7	0	0	0	0
2	1	0	0	0	0	0
3	0.5	0	0	0	0	0.5
4	0	0	0.6	0	0	0.4
5	0	0	0	0	0.1	0.9
6	0	0	0	0	0.7	0.3

The diagram for this chain is given below. The corresponding R code will be presented in Example 1.9.

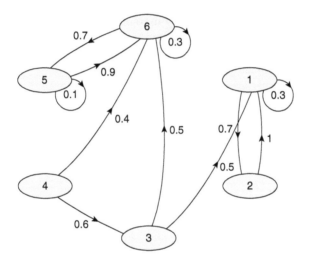

We can see that states 3 and 4 are reflecting and therefore transient. The chain will transition out of both states and will enter the recurrent class $\{1, 2\}$ or $\{5, 6\}$. States 3 and 4 have infinite periods whereas both recurrent classes are aperiodic due to the existence of the loops. □

1.5 Limiting Probabilities

In a Markov chain, the *limiting probability* $\pi_j = \lim_{n \to \infty} P_{ij}^n$ doesn't depend on the initial state i and can be interpreted as the long-run proportion of time that the Markov chain spends in state j. Limiting probabilities are also termed the *limiting distribution* or *stationary distribution* or *steady-state distribution*. If the stationary distribution exists, it satisfies the system of equations:

$$\pi_j = \sum_{i=0}^{\infty} \pi_i P_{ij}. \tag{1.2}$$

To see this, we condition on the state at time n and write $\mathbb{P}(X_{n+1} = j) = \sum_{i=0}^{\infty} \mathbb{P}(X_{n+1} = j | X_n = i) \mathbb{P}(X_n = i) = \sum_{i=0}^{\infty} P_{ij} \mathbb{P}(X_n = i)$. Letting $n \to \infty$, and assuming that we can justify exchanging the limit and the summation sign, we get (1.2).

If we combine the limiting probabilities into a row vector $\pi = (\pi_1, \pi_2, \dots)$, then the system of equations (1.2) has the matrix form $\pi = \pi \cdot \mathbf{P}$ where $\mathbf{P}$ is the one-step transition probability matrix. Since the rows of the transition probability matrix add up to 1, the equations in the system are linearly dependent. Nonetheless, we can find all limiting probabilities if we take into account the fact that they must sum up to 1: $\pi_1 + \pi_2 + \dots = 1$.

A Markov chain that has a unique stationary distribution is referred to as *ergodic*.

EXAMPLE 1.6. The stationary distribution for the Markov chain considered in Examples 1.1 – 1.4 solves

$$(\pi_1, \pi_2, \pi_3) = (\pi_1, \pi_2, \pi_3) \begin{bmatrix} 0.7 & 0.1 & 0.2 \\ 0.0 & 0.6 & 0.4 \\ 0.5 & 0.2 & 0.3 \end{bmatrix}, \quad \text{and} \quad \pi_1 + \pi_2 + \pi_3 = 1.$$

Or, equivalently,

$$\begin{cases} \pi_1 = 0.7\pi_1 + 0.5\pi_3 \\ \pi_2 = 0.1\pi_1 + 0.6\pi_2 + 0.2\pi_3 , \\ \pi_1 + \pi_2 + \pi_3 = 1, \end{cases} \quad \text{or} \quad \begin{cases} \pi_3 = 0.6\pi_1 \\ \pi_2 = 0.55\pi_1 \\ 2.15\pi_1 = 1, \end{cases}$$

resulting in the solution $\pi_1 = 0.4651$, $\pi_2 = 0.2558$, and $\pi_3 = 0.2791$. It means that in a long run, the chain spends roughly 46.5% of the time in state 1, 25.6% of the time in state 2, and 27.9% of the time in state 3. Since the stationary distribution is unique, it is an ergodic chain. $\square$

EXAMPLE 1.7. As will be demonstrated in Example 1.9, for the Markov chain in Example 1.5, the system of equations (1.2) has two solutions

$$(0.5882, 0.4118, 0, 0, 0, 0) \quad \text{and} \quad (0, 0, 0, 0, 0.4375, 0.5625).$$

This happens because the chain contains two recurrent classes, $\{1, 2\}$ and $\{5, 6\}$. Since there are two solutions, the chain is *non-ergodic*. Denote these two vectors by π_1 and π_2, respectively. Then any linear combination of the form $\alpha\pi_1 + (1-\alpha)\pi_2$ where $0 < \alpha < 1$ is a solution, and therefore, there actually exist infinitely many solutions. Any such solution is called an *invariant measure*. Neither is considered to be the stationary distribution. □

1.6 Computations in R

To mimic in R the analysis done in examples in the previous section, three packages are required: "diagram" to plot the diagram, "expm" to exponentiate matrices, and "markovchain" to determine recurrent and transient classes, absorbing states, and the steady-state distribution. The script is as follows:

```
install.packages("diagram")
install.packages("expm")
install.packages("markovchain")
```

• First, an $n \times n$ transition probability matrix should be specified:

tm.name<- matrix(c(p_{11}.*value*, p_{12}.*value*, ..., p_{nn}.*value*), nrow=*n.value*, ncol=*n.value*, byrow=TRUE)

• To plot the diagram with the arrows pointing in the correct direction, the transition matrix must be transposed. This can be done by typing

tr.tm.name<- t(*tm.name*)

The diagram is drawn using plotmat function. The graph depicts state names inside boxes and directed lines connecting the boxes labeled by the corresponding transition probabilities. The syntax is:

```
library(diagram)
plotmat(tr.tm.name, <arguments>)
```

The arguments in the above function are:

○ pos, a vector specifying the number of boxes in each row in the diagram. For instance, in a three-state chain, pos=c(1,2) means that state 1 is depicted in the top row, and states 2 and 3 are in the next row. By default, boxes are plotted in a circle.
○ name=c("*state1.name*", "*state2.name*", ...), the list of state names. By default, natural numbers are used.
○ arr.col, color of inside of all arrowheads (excluding the contours). Black by default.

○ `arr.lcol`, color of all arrow lines. Black by default.

○ `arr.length`, length of all arrows.

○ `arr.pos`, relative position of arrowheads on lines (excluding loops), a value between 0 and 1. By default, arrows are positioned in the middle.

○ `arr.type`, type of arrowhead, some options are `curved`, `triangle`, or `simple`.

○ `arr.width`, width of all arrows.

○ `box.cex`, size of labels in boxes (i.e., state names). The magnitude is relative to the default value of character expansion.

○ `box.col`, color of inside of all boxes (excluding contours). By default, the color is white.

○ `box.lcol`, color of contours of all boxes. Black by default.

○ `box.lwd`, width of contours of all boxes.

○ `box.prop`, ratio of length over width of all boxes. The ratio is equal to 1 by default.

○ `box.size`, size of all boxes.

○ `box.type`, type of all boxes, some options are `rect`, `ellipse`, `round` (a rectangle with rounded edges), `circle`, `diamond`, `hexa` (a hexagonal shape).

○ `cex.txt`, size of labels next to arrows (i.e., values of respective probabilities).

○ `lcol`, color of all lines, contours of all arrows, and contours of all boxes. Black by default.

○ `lwd`, width of all arrow lines, excluding loops.

○ `self.cex`, size of all loops (also termed "self-arrows").

○ `self.arrpos`, angle in radians of arrow positions on all loops relative to the x-axis.

○ `self.lwd`, width of all loops.

○ `self.shiftx`, shift of all loops along the x-axis.

○ `self.shifty` shift of all loops along the y-axis.

○ `txt.col`, color of labels in boxes (i.e., state names).

• To compute an n-step transition probability matrix for a specific value of n, use the following code:

library(expm)
nstepmatrix.name<- *tm.name*% $\wedge$ %*n.value*

• Given the initial distribution, the lines below calculate the unconditional distribution after n steps.

init.p.name <- c(*p1.init.value, p2.init.value, . . .*)
uncond.dist.name<- *init.p.name*%*%nstepmatrix.name*

• To determine recurrent and transient classes, absorbing states, and the limiting distribution, a discrete-time Markov chain must be created as an object. It can be done as follows:

```
library(markovchain)
dtmc.name<- new("markovchain", transitionMatrix=tm.name,
states=c("state1.name", "state2.name",...))
```

Then state characteristics may be obtained as

```
recurrentClasses(dtmc.name)
transientClasses(dtmc.name)
absorbingStates(dtmc.name)
steadyStates(dtmc.name)
```

Note that R doesn't identify reflecting states.

In R, one can compute the period of an irreducible (a single-class) Markov chain only. The syntax is `period(dtmc.name)`.

EXAMPLE 1.8. The results obtained for the Markov chain from Examples 1.1-1.4 and 1.6 can be produced in R as presented below.

```
#specifying transition probability matrix
tm<- matrix(c(0.7, 0.1, 0.2, 0.0, 0.6, 0.4, 0.5, 0.2, 0.3),
nrow=3, ncol=3, byrow=TRUE)

#transposing transition probability matrix
tm.tr<- t(tm)

#plotting diagram
library(diagram)
plotmat(tm.tr, pos=c(1,2), arr.length=0.3, arr.width=0.1,
box.col="light blue", box.lwd=1, box.prop=0.5, box.size=0.12,
box.type="circle", cex.txt=0.8, lwd=1, self.cex=0.6,
self.shiftx=0.17, self.shifty=-0.01)
```

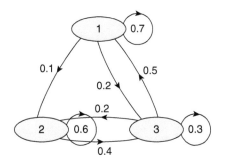

```
#computing three-step transition probability matrix
library(expm)
print(tm3<- tm%∧% 3)
```

```
      [,1]  [,2]  [,3]
[1,] 0.533 0.209 0.258
[2,] 0.320 0.356 0.324
[3,] 0.485 0.242 0.273
```

```
#computing unconditional distribution after three steps
init.p<- c(1/3, 1/3, 1/3)
init.p%*%tm3
```

```
      [,1]  [,2]  [,3]
[1,] 0.446 0.269 0.285
```

```
#creating Markov chain object
library(markovchain)
mc<- new("markovchain", transitionMatrix=tm, states=c("1",
"2", "3"))
```

```
#computing Markov chain characteristics
recurrentClasses(mc)
```

```
"1" "2" "3"
```

```
transientClasses(mc)
```

```
list()
```

```
absorbingStates(mc)
```

```
character(0)
```

```
period(mc)
```

```
1
```

```
round(steadyStates(mc), digits=4)
```

```
     1      2      3
0.4651 0.2558 0.2791
```

□

EXAMPLE 1.9. Consider the Markov chain from Examples 1.5 and 1.7. We run the following R code to produce the diagram, invariant vectors, and to verify the state classification.

```
#specifying transition probability matrix
tm<- matrix(c(0.3,0.7,0,0,0,0,1,0,0,0,0,0,0.5,0,0,0,0,0.5,
0,0,0.6,0,0,0.4,0,0,0,0,0.1,0.9,0,0,0,0,0.7,0.3), nrow=6,
ncol=6, byrow=TRUE)

#transposing transition probability matrix
tm.tr<- t(tm)

#plotting diagram
library(diagram)
plotmat(tm.tr, arr.length=0.3, arr.width=0.1, box.col="light
blue", box.lwd=1, box.prop=0.5, box.size=0.09,
box.type="circle", cex.txt=0.8, lwd=1, self.cex=0.3,
self.arrpos=0.3, self.shiftx=0.09, self.shifty=-0.05)
```

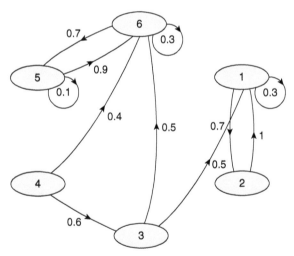

```
#creating Markov chain object
library(markovchain)
mc<- new("markovchain", transitionMatrix=tm, states=c("1",
"2", "3", "4", "5", "6"))

#computing Markov chain characteristics
recurrentClasses(mc)
```

"1" "2"

"5" "6"

```
transientClasses(mc)
```

"3"

"4"

```
#finding periods of irreducible Markov chains
tm12.ir<- matrix(c(0.3,0.7,1,0), nrow=2, ncol=2, byrow=TRUE)
mc12.ir<- new("markovchain", transitionMatrix=tm12.ir,
states=c("1","2"))
period(mc12.ir)
```

1

```
tm56.ir<- matrix(c(0.1,0.9,0.7,0.3), nrow=2, ncol=2,
byrow=TRUE)
mc56.ir<- new("markovchain", transitionMatrix=tm56.ir,
states=c("5","6"))
period(mc56.ir)
```

1

```
#finding steady-state distribution
round(steadyStates(mc), digits=4)
```

1	2	3	4	5	6
0.0000	0.0000	0	0	0.4375	0.5625
0.5882	0.4118	0	0	0.0000	0.0000

□

1.7   Simulations in R

In R, to simulate simultaneously *ntraj.value* trajectories of a k-state Markov chain over *nsteps.name* time-steps, the function **rmarkovchain()** in the library **markovchain** can be used. It keeps track of the step number and the Markov chain state at that step. Each new state is chosen according to a multinomial probability distribution with the probability mass function

$$P(x_1, \cdots, x_k, p_1, \cdots, p_k) = \mathbb{P}(X_1 = x_1, \cdots, X_k = x_k)$$

$$= \frac{n!}{x_1! \cdots \cdots x_k!} p_1^{x_1} \cdots \cdots p_k^{x_k}, \text{ where } x_1 + \cdots + x_k = n.$$

Given the initial vector of probabilities $(p_1^1, \ldots, p_k^1)$, the multivariate probability distribution is found recursively:

$$(p_1^n, p_2^n, \ldots, p_k^n) = (p_1^{n-1}, p_2^{n-1}, \ldots, p_k^{n-1}) \, \mathbf{P}, \quad n = 2, 3, \ldots, \qquad (1.3)$$

where $\mathbf{P}$ denotes the one-step transition probability matrix of the Markov chain.

Assuming that the Markov chain object *mc.name* has already been created, the syntax is as follows:

```
#specifying total number of trajectories
ntraj.name<- ntraj.value

#specifying total number of steps
nsteps.name<- nsteps.value

#specifying initial probability
init.prob.name<- c(p1.value, p2.value, ..., pk.value)

#specifying matrix containing states
states.matrix.name<- matrix(NA, nrow=nsteps.name, ncol=ntraj.name)

#specifying seed
set.seed(value)

#generating initial state
init.state.name<- sample(1:k, 1, prob=init.prob.name)

#simulating states
for (i in 1:ntraj.name)
states.matrix.name[,i]<- rmarkovchain(n=nsteps.name-1,
object=mc.name,
t0=init.state.name, include.t0=TRUE)
```

• The value of the seed tells R where to start reading off in the table of random digits. This is done for the reproducibility of results. If the code is run again, it will produce the same trajectories.

• The function `sample(1:k, 1, prob=`*init.prob.name*`)` draws one state from among the k states and the sampling is done according to the multinomial distribution with the probability vector *init.prob.name*.

• The final product of the above code is a matrix with dimensions *nsteps.value* by *ntraj.value* with columns containing state names for individual trajectories.

As an alternative to using the built-in function `rmarkovchain()`, one can create a user-defined function that simulates trajectories. The syntax is

```
function.name<- function(tm.name, init.prob.name, nsteps.name) {
states.name<- numeric(nsteps.name)
states.name[1]<- sample(1:k, 1, prob=init.prob.name)

   for(t in 2:nsteps.name) {
prob.name<- tm.name[states.name[t-1],]
   states.name[t]<- sample(1:k, 1, prob=prob.name)
   }
   return(states.name)
   }
```

```
set.seed(value)
states.matrix.name<- matrix(NA, nrow=nsteps.name, ncol=ntraj.name)
   for (j in 1:ntraj.name)
states.matrix.name<- function.name(tm.name, init.prob.name,
   nsteps.name)
```

Finally, the *ntraj.value* trajectories can be overlayed on a single graph by means of the `matplot()` function with the following syntax:

```
#plotting simulated trajectories
matplot(states.matrix.name, <arguments>)
```

where the arguments are: `type="l"` to connect the dots, `lty=1` to draw a solid line, `lwd=2` to make the lines appear thicker, `col=a:b` to choose a sequential subset *a* to *b* of *ntraj.name* colors from this cyclic list {1=black, 2=red, 3=green, 4=dark blue, 5=sky blue, 6=pink, 7=yellow, 8=gray, 9=black, 10=red, 11=green, etc.}, `xlim=c(`*x.lower.value*, *x.upper.value*`)` to limit the *x*-axis, if needed, `ylim=c(`*y.lower.value*, *y.upper.value*`)` to limit the *y*-axis, if needed, `xlab="`*step.name*`"` and `ylab="`*state.name*`"` to display *step.name* on the *x*-axis and *state.name* on the *y*-axis. Colors can also be specified as an explicit list: `col=c("red", "green", "blue")`, for example. To add grid lines to the coordinate system on the plot, the argument `panel.first=grid()` is used.

To change tick marks on either axis and to modify labels for ticks, one can remove axes by including in `matplot()` the arguments `xaxt="n"` and `yaxt="n"`, and then specify the new axes as `axis(side=`*number*, `at=`*range for ticks*`)`, where `side=1` is for *x*-axis, and `side=2` is for *y*-axis.

To enhance the clarity of what states the depicted trajectories are at, dots can be added at every step. To do so, use `points(`*x values*`, `*y values*`, pch=16)`, where the argument `pch=16` identifies the plotting character for the points as a filled circle.

Additionally, to visualize convergence of unconditional probabilities p_n, computed recursively in accordance with (1.3), the following R syntax can be implemented. We assume that there are k states in the Markov chain, and the transition probability matrix has been defined as *tm.name* object.

```
#specifying total number of steps
nsteps.name<- nsteps.value

#specifying matrix containing probabilities
probs.matrix.name<- matrix(NA, nrow=nsteps.name, ncol=k)

#specifying initial probability
init.prob.name<- c(p1.value, p2.value, ..., pk.value)

#computing unconditional probabilities
probs.matrix.name[1,] <- init.prob.name
   for(n in 2:nsteps.name)
   probs.matrix.name[n,]<- probs.matrix.name[n-1,]%*%tm.name

#plotting probabilities against steps by state
matplot(probs.matrix.name, type="l", lty=1, col=a:b,
ylim=c(lower.value,upper.value), ylab="probability", xlab="step")
```

In addition, a legend should be added to match the lines with states' numbers.

```
legend(legend.position, c("state 1", "state 2",..., "state k"),
lty=1, col=a:b)
```

- The choices for `legend.position` on the graph are: `"topright"`, `"right"`, `"bottomright"`, `"top"`, `"center"`, `"bottom"`, `"topleft"`, `"left"`, and `"bottomleft"`.

SIMULATION 1.1. Consider the Markov chain discussed in Examples 1.1-1.4, 1.6, and 1.8. Recall that the transition probability matrix for this chain is specified as

$$\begin{bmatrix} 0.7 & 0.1 & 0.2 \\ 0.0 & 0.6 & 0.4 \\ 0.5 & 0.2 & 0.3 \end{bmatrix}.$$

Below we simulate two trajectories of this chain with the initial state chosen at random, that is, with the probability vector $(1/3, 1/3, 1/3)$. First, we demonstrate how to utilize the built-in function rmarkovchain().

```
#specifying transition probability matrix
tm<- matrix(c(0.7, 0.1, 0.2, 0.0, 0.6, 0.4, 0.5, 0.2, 0.3),
nrow=3, ncol=3, byrow=TRUE)

#creating Markov chain object
library(markovchain)
mc<- new("markovchain", transitionMatrix=tm, states=c("1",
"2", "3"))

#specifying total number of steps
nsteps<- 25

#specifying initial probability
p0<- c(1/3, 1/3, 1/3)

#specifying matrix containing states
MC.states<- matrix(NA, nrow=nsteps, ncol=2)

#specifying seed
set.seed(2443927)

#simulating trajectories
for (i in 1:2)
state0<- sample(1:3, 1, prob=p0)
MC.states[,i]<- rmarkovchain(n=nsteps-1, object=mc,
t0=state0, include.t0=TRUE)

#plotting simulated trajectories
matplot(MC.states, type="l", lty=1, lwd=2, col=3:4,
xaxt="n", yaxt="n", ylim=c(1,3), xlab="Step", ylab="State",
panel.first=grid())

axis(side=1, at=c(1,5,10,15,20,25))
axis(side=2, at=1:3)

points(1:nsteps, MC.states[,1], pch=16, col=3)
points(1:nsteps, MC.states[,2], pch=16, col=4)
```

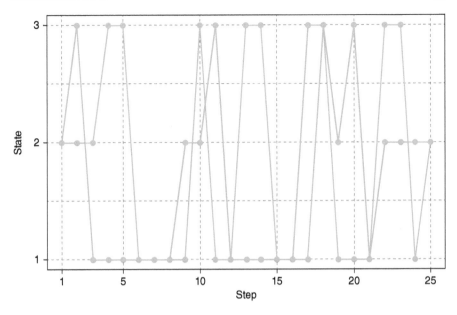

Now we present the code with a user-defined function that simulates trajectories, bypassing the built-in **rmarkovchain()** function. We use the same seed as above and obtain the same trajectories. The code and graph follow.

```
#creating user-defined function
MC<- function(tm, p0, nsteps) {
states<- numeric(nsteps)
states[1]<- sample(1,1,prob=p0)

for(t in 2:nsteps) {
  p<- tm[states[t-1],]
    states[t]<- sample(1,1,prob=p)
}
    return(states)
}

#specifying seed
set.seed(2443927)

#simulating trajectories
MC.states2<- matrix(NA, nrow=nsteps, ncol=2)
for (j in 1:2)
MC.states2[,j]<- MC(tm, p0, nsteps)
```

```
#plotting simulated trajectories
matplot(MC.states2, type="l", lty=1, lwd=2, col=3:4,
xaxt="n", yaxt="n", ylim=c(1,3), xlab="Step", ylab="State",
panel.first= grid())

axis(side=1, at=c(1,5,10,15,20,25))
axis(side=2, at=c(1,2,3))

points(1:nsteps, MC.states2[,1], pch=16, col=3)
points(1:nsteps, MC.states2[,2], pch=16, col=4)
```

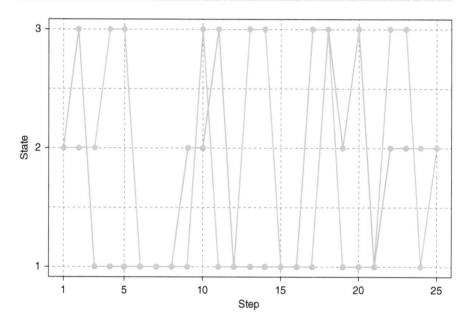

Now we compute iteratively probability vectors and plot them against the step number. The code and plot are given below.

```
#specifying matrix containing probabilities
probs<- matrix(NA, nrow=nsteps, ncol=3)

#specifying total number of steps
nsteps<- 15
```

```
#computing probabilities p_n
probs[1,] <- p0
for(n in 2:nsteps)
probs[n,]<- probs[n-1,]%*%tm

#plotting probabilities against steps by state
matplot(probs, type="l", lty=1, lwd=2, col=2:4,
ylim=c(0.2,0.5), xlab="Step ", ylab="Probability",
panel.first=grid())
legend("right", c("State 1 ", "State 2", "State 3"), lty=1,
lwd=2, col=2:4)
```

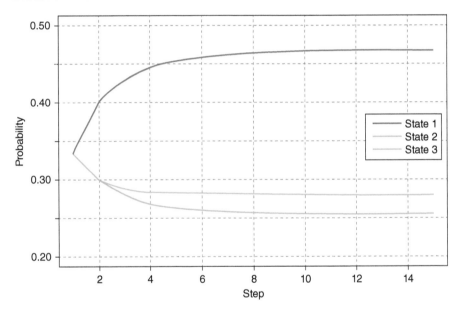

We also output the values of the probabilities to see that they converge to the steady state $(0.4651, 0.2558, 0.2791)$ on step 14.

```
> probs
              [,1]       [,2]       [,3]
              <lines omitted>
[13,] 0.4650353 0.2558698 0.2790949
[14,] 0.4650721 0.2558444 0.2790835
```

□

SIMULATION 1.2. We return to the Markov chain studied in Examples 1.5, 1.7, and 1.9. First, we use the already-created object mc to generate three trajectories that start at a randomly chosen state. In the code that follows, we apply the **rmarkovchain()** function to simulate the trajectories.

```
#specifying total number of steps
nsteps<- 20

#specifying initial probability
p0<- c(1/6, 1/6, 1/6, 1/6, 1/6, 1/6)

#specifying matrix containing states
MC.states<- matrix(NA, nrow=nsteps, ncol=3)

#specifying seed
set.seed(765881)

#simulating trajectories
for (i in 1:3) {
state0<- sample(1:6, 1, prob=p0)
MC.states[,i]<- rmarkovchain(n=nsteps-1, object=mc,
t0=state0,
 include.t0=TRUE)
}

#plotting simulated trajectories
matplot(MC.states, type="l", lty=1, lwd=2, col=2:4, xaxt="n",
ylim=c(1,6), xlab="Step", ylab="State", panel.first=grid())

axis(side=1, at=c(1,5,10,15,20))

points(1:nsteps, MC.states[,1], pch=16, col=2)
points(1:nsteps, MC.states[,2], pch=16, col=3)
points(1:nsteps, MC.states[,3], pch=16, col=4)
```

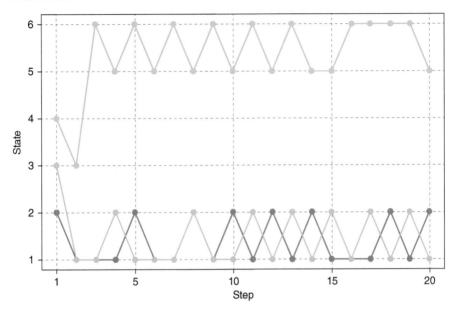

Next, we compute and plot the unconditional probabilities p_n against n. As n increases, these probabilities converge to an invariant probability measure, which heavily depends on the initial state of the Markov chain. These limiting values will be the same for states 1 and 2 since they belong to the same class, for states 5 and 6 for the same reason, and separately for state 3, and for state 4. From the theoretical viewpoint, the invariant measure for states 1 and 2 is $(0.5882, 0.4118, 0, 0, 0, 0)$. For states 5 and 6, it is $(0, 0, 0, 0, 0.4375, 0.5625)$. From state 3 the chain is equally likely to enter either $\{1, 2\}$ or $\{5, 6\}$, therefore, the invariant measure is found as $(0.5)(0.5882, 0.4118, 0, 0, 0, 0) + (0.5)(0, 0, 0, 0, 0.4375, 0.5625) = (0.2941, 0.2059, 0, 0, 0.21875, 0.28125)$. From state 4, the chain will enter the class $\{5, 6\}$ directly with probability 0.4, or through state 3, with probability $(0.6)(0.5) = 0.3$, hence, it enters class $\{5, 6\}$ with the total probability $0.4 + 0.3 = 0.7$, and enters class $\{1, 2\}$ only through state 3 with the probability $(0.6)(0.5) = 0.3$. As a result, the invariant measure for state 4 is $(0.3)(0.5882, 0.4118, 0, 0, 0, 0) + (0.7)(0, 0, 0, 0, 0.4375, 0.5625) = (0.17646, 0.12354, 0, 0, 0.30625, 0.39375)$.

We run the code below six times, every time choosing a different initial state. As the output, we present the graphs for each initial state, and print the vector of probabilities for steps 28 through 30.

```
#specifying total number of steps
nsteps<- 30
```

```
#specifying initial state distribution (state 1)
p0<- c(1,0,0,0,0,0)
#(state 2) p0<- c(0,1,0,0,0,0), (state 3) p0<-
c(0,0,1,0,0,0),
#(state 4) p0<- c(0,0,0,1,0,0), (state 5) p0<-
c(0,0,0,0,1,0),
#(state 6) p0<- c(0,0,0,0,0,1)

#specifying matrix containing probabilities
probs<- matrix(NA, nrow=nsteps, ncol=6)

#computing probabilities p_n
probs[1,] <- p0

for(n in 2:nsteps)
probs[n,]<- probs[n-1,]%*%tm

#plotting probabilities vs. step by state
matplot(probs, main="Initial State 1", type="l", lty=1,
lwd=2, col=1:6, ylim=c(-0.05, 1.1), xlab="Step",
ylab="Probability", panel.first = grid())

legend("topright", c("State 1", "State 2", "State 3", "State
4", "State 5", "State 6"), lty=1, lwd=2, col=1:6)
```

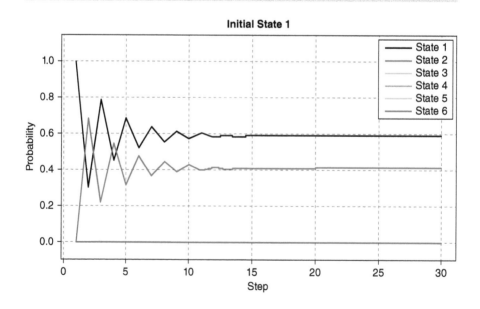

probs

	[,1]	[,2]	[,3]	[,4]	[,5]	[,6]
		<lines omitted>				
[28,]	0.5882082	0.4117918	0	0	0	0
[29,]	0.5882542	0.4117458	0	0	0	0
[30,]	0.5882220	0.4117780	0	0	0	0

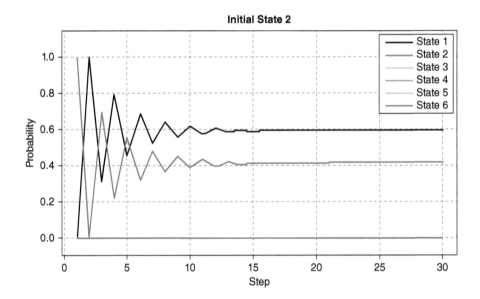

	[,1]	[,2]	[,3]	[,4]	[,5]	[,6]
		<lines omitted>				
[28,]	0.5882739	0.4117261	0	0	0	0
[29,]	0.5882082	0.4117918	0	0	0	0
[30,]	0.5882542	0.4117458	0	0	0	0

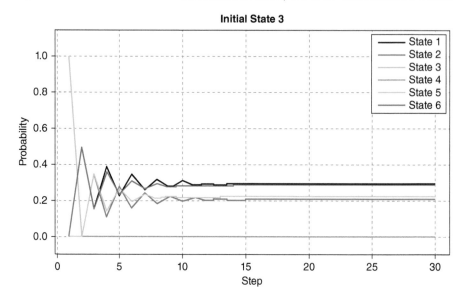

	[,1]	[,2]	[,3]	[,4]	[,5]	[,6]
		<lines omitted>				
[28,]	0.2941370	0.2058630	0	0	0.2187496	0.2812504
[29,]	0.2941041	0.2058959	0	0	0.2187502	0.2812498
[30,]	0.2941271	0.2058729	0	0	0.2187499	0.2812501

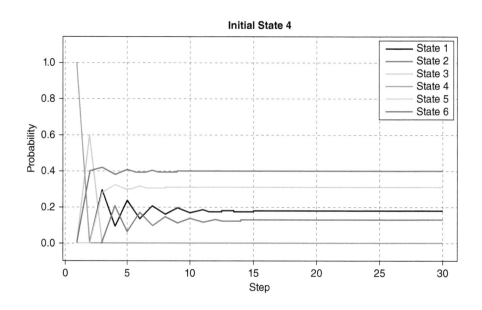

	[,1]	[,2]	[,3]	[,4]	[,5]	[,6]
		<lines omitted>				
[28,]	0.1764540	0.1235460	0	0	0.3062501	0.3937499
[29,]	0.1764822	0.1235178	0	0	0.3062500	0.3937500
[30,]	0.1764625	0.1235375	0	0	0.3062500	0.3937500

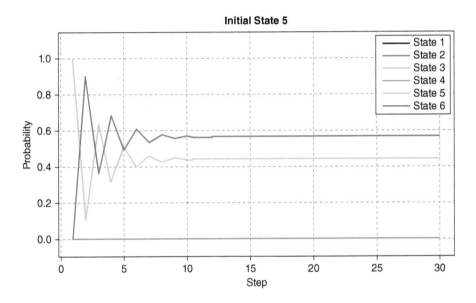

	[,1]	[,2]	[,3]	[,4]	[,5]	[,6]
		<lines omitted>				
[28,]	0	0	0	0	0.4374994	0.5625006
[29,]	0	0	0	0	0.4375003	0.5624997
[30,]	0	0	0	0	0.4374998	0.5625002

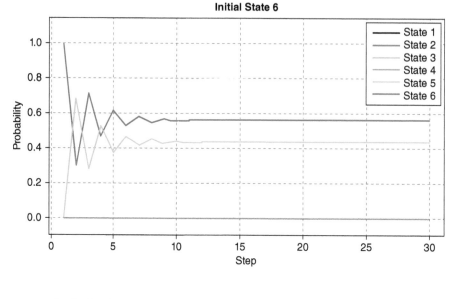

	[,1]	[,2]	[,3]	[,4]	[,5]	[,6]
				<lines omitted>		
[28,]	0	0	0	0	0.4374994	0.5625006
[29,]	0	0	0	0	0.4375003	0.5624997
[30,]	0	0	0	0	0.4374998	0.5625002

□

1.8   Applications of Markov Chain

APPLICATION 1.1. In 1913, A. A. Markov (1856-1922), a Russian mathematician after whom Markov chains are named, published an article where he analyzed sequences of vowels and consonants among the first 20,000 letters of "Eugene Onegin" by A. S. Pushkin.[2] Two silent letters in the Russian language that are indicators of the softness of a preceding sound and are neither vowels nor consonants were ignored. He cleverly argued that the appearance of vowels and consonants are remarkably dependent in such a way that this literary work may be approximated by what we now call a Markov chain. He went ahead and computed the one-step transition probability matrix for this chain with the state space $S = \{vowel\,(\text{v}), consonant\,(\text{c})\}$. He calculated that

---

[2]Markov, A. A. (1913). "An example of statistical investigation of the text Eugene Onegin concerning the connection of samples in chains." (In Russian). *Bulletin of the Imperial Academy of Sciences of St. Petersburg*, 7(3): 153 – 162.

there are $8,638$ vowels in the text (respectively, $11,362$ consonants) and $1,104$ vowel-vowel sequences. It is also important to notice that the text starts with a consonant and ends with a vowel, meaning that nothing transitions into the first consonant and the total number of transitions into vowels adds up to the total number of vowels. Thus, there must be $8,638-1,104 = 7,534$ consonant-vowel sequences, $11,362-7,534 = 3,828$ consonant-consonant sequences, and $11,361-3,828 = 7,533$ vowel-consonant sequences. The transition probability matrix can then be computed as

$$\mathbf{P} = \begin{array}{c} \\ v \\ \\ c \end{array} \begin{bmatrix} \overset{v}{\dfrac{1104}{8637}} = 0.1278 & \overset{c}{\dfrac{7533}{8637}} = 0.8722 \\ \dfrac{7534}{11362} = 0.6631 & \dfrac{3828}{11362} = 0.3369 \end{bmatrix}.$$

Now, out of curiosity, we compute the limiting probabilities, which, theoretically speaking, represent proportions of vowels and consonants in the text. Since we know the exact numbers, we compute $\pi_v = \frac{8638}{20000} = 0.4319$, and $\pi_c = 1 - 0.4319 = 0.5681$. Now, resorting to R, we obtain the same values:

```
#specifying the transition probability matrix
tm<- matrix(c(0.1278, 0.8722, 0.6631, 0.3369), nrow=2,
ncol=2,
byrow=TRUE)

#creating Markov chain object
library(markovchain)
mc<- new("markovchain", transitionMatrix=tm, states=c("v",
"c"))
#computing limiting probabilities
steadyStates(mc)
```

```
        v         c
0.4319026 0.5680974
```

To verify how accurate A. A. Markov was with his estimation of the one-step transition probability matrix, we ran the same analysis on the entire novel. First, we precleaned the text, removing everything besides the Cyrillic letters of the pre-1918 reform Russian alphabet. Then we ran the code presented below that changes the capitalization to lowercase, then removes blanks, line breaks, punctuation marks, and the soft and hard signs of the Russian alphabet, leaving all the cleaned analysis-ready string containing a total of $106,508$ characters. To compute the combinations vv, cv, vc, and cc, we shift the string to the left by one position, by inserting a blank in the front and deleting the last letter to preserve the length of the string. When we line up the lagged string x_1 with the original string x_2, we obtain the preceding and following letters in the string. After that, we apply simple Boolean logic to calculate the number of vowels and consonants and the number of combinations of all four types. The code and output follow.

```
text <- read_file("./Onegin.txt")

#text cleaning
#gsub() = global substitution function=replaces all instances
lowercase<- tolower(text)
no.blanks<- gsub(""," ",lowercase)
no.line.breaks<- gsub("\r\n", "", no.blanks)
no.punctuation<- gsub("[[:punct:]]","",no.line.breaks)
no.soft.signs<- gsub("ь", "", no.punctuation)
#removing all hard signs
clean.string<- gsub("ъ","", no.soft.signs)
```

```
#splitting single string into characters
x2<- strsplit(clean.string, "")

#shifting string by one position
no.last<- substr(clean.string, 1, nchar(clean.string)-1)
first.blank<- str_c("", no.last)
x1<- strsplit(first.blank,"")

#Note: In pre-1918 Russian language "й" was considered a
vowel
vowels<-c("a", "e", "ё", "и", "i", "й", "o", "y", "ы", "ѣ",
"э", "ю", "я")

consonants<- c("б", "в", "г", "д", "ж", "з", "к", "л", "м",
"н", "п", "р", "с", "т", "ф", "х", "ц", "ч", "ш", "щ", "ѳ")

#computing number of vowels, consonants, and four
combinations
for (counter in 1:nchar(x2)){
v<- ifelse(x2[[counter]] %in% vowels,1,0)
c<- ifelse(x2[[counter]] %in% consonants,1,0)
```

```
vv<- ifelse(x1[[counter]] %in% vowels & x2[[counter]] %in%
vowels,1,0)
vc<- ifelse(x1[[counter]] %in% vowels & x2[[counter]] %in%
consonants,1,0)
cv<- ifelse(x1[[counter]] %in% consonants & x2[[counter]]
%in% vowels,1,0)
cc<- ifelse(x1[[counter]] %in% consonants & x2[[counter]]
%in% consonants,1,0)
}
sum(v)
```

46475

```
sum(c)
```

60033

```
sum(vv)
```

6368

```
sum(vc)
```

40107

```
sum(cv)
```

40107

```
sum(cc)
```

19925

Note that all these numbers add up perfectly. The total number of vowels and consonants is $46,475 + 60,033 = 106,508$. Since the text starts and ends with consonants, all vowels have leading and trailing letters and therefore, we must have sum(v)=sum(vv)+sum(vc)=sum(vv)+sum(cv). Indeed, sum(vc)=sum(cv)=40,107 and sum(vv)+sum(vc) = $6,368 + 40,107 = 46,475$ =sum(v). Also, all consonants but the first one have leading letters and all consonants but the last one have trailing letters. Hence, we must have sum(c)-1=sum(cv)+sum(cc)=sum(vc)+sum(cc), which indeed holds since $40,107 + 19,925 = 60,032 = 60,033 - 1$.

Turning these numbers into the one-step transition probability matrix, we obtain

$$
\mathbf{P} = \begin{array}{c} \\ v \\ c \end{array}
\begin{bmatrix}
\dfrac{6368}{46475} = 0.1370 & \dfrac{40107}{46475} = 0.8630 \\[2ex]
\dfrac{40107}{60032} = 0.6681 & \dfrac{19925}{60032} = 0.3319
\end{bmatrix}.
$$

Note that A. A. Markov used about one-fifth of the text and obtained pretty accurate estimates. □

APPLICATION 1.2. It is important to understand that it is virtually impossible to show that a given process is a Markov chain because we would need to show that the Markovian property (1.1) holds for all tuples: pairs, triples, quadruples, quintuples, etc. of successive observations. In his 1913 article (see the previous application), A. A. Markov went only as deep as looking at triples, and thus he could only conclude that the process is approximately Markovian (put in modern terms).

On the other hand, showing that a given process is non-Markovian is much easier because it suffices to show that the definition (1.1) fails for some tuple. For instance, if we can show that for some fixed $i_1, i_2,$ and i_3, $\hat{\mathbb{P}}(X_3 = i_3 \,|\, X_1 = i_1, \; X_2 = i_2) \neq \hat{\mathbb{P}}(X_3 = i_3 \,|\, X_2 = i_2)$, it would be enough to prove that the process is not a Markov chain. Below we give an example of such a process.

Staying on the topic of literature, recall that in the famous 1980 movie "Shining," the main character Jack Torrance (played by Jack Nicholson) repeatedly typed on a typewriter the sentence "All work and no play makes Jack a dull boy." Suppose he typed it up k times, where k is a very large number. If we turn this "composition" into a sequence of vowels (v) (a, e, i, o, u), and consonants (c) (the remaining 21 letters), we will get a truly deterministic sequence with a repeating pattern:

```
allworkandnoplaymakesjackadullboy|allworkandnoplay...
vcccvccvcccvccvccvcvccvccvcvcccvc|vcccvccvcccvccvc...
```

We want to show that the Markovian property doesn't hold, for instance, for the triples {ccc} and thus this chain is non-Markov. In the entire "composition" there are $33k$ letters, of which $11k$ are vowels and $22k$ are consonants. Also, pattern {ccc} occurs $3k$ times, {ccv} occurs $8k$ times, {vcc} occurs $8k$ times, and {vcv} occurs $2k + k - 1 = 3k - 1$ times because two such triples lie inside each sentence and one appears at each seam. Thus, we find

$$\hat{\mathbb{P}}(X_3 = \mathsf{c} \,|\, X_2 = \mathsf{c}, \; X_1 = \mathsf{c}) = \frac{\hat{\mathbb{P}}(X_1 = \mathsf{c}, \; X_2 = \mathsf{c}, \; X_3 = \mathsf{c})}{\hat{\mathbb{P}}(X_1 = \mathsf{c}, \; X_2 = \mathsf{c})}$$

$$= \frac{\hat{\mathbb{P}}(X_1 = \mathsf{c}, \; X_2 = \mathsf{c}, \; X_3 = \mathsf{c})}{\hat{\mathbb{P}}(X_1 = \mathsf{c}, \; X_2 = \mathsf{c}, \; X_3 = \mathsf{v}) + \hat{\mathbb{P}}(X_1 = \mathsf{c}, \; X_2 = \mathsf{c}, \; X_3 = \mathsf{c})}$$

$$= \frac{\hat{\mathbb{P}}(\mathsf{ccc})}{\hat{\mathbb{P}}(\mathsf{ccv}) + \hat{\mathbb{P}}(\mathsf{ccc})} = \frac{3k}{8k + 3k} = \frac{3}{11}.$$

On the other hand, we estimate

$$\widehat{\mathbb{P}}\left(X_3 = \mathsf{c} \mid X_2 = \mathsf{c}\right) = \frac{\widehat{\mathbb{P}}\left(X_2 = \mathsf{c},\, X_3 = \mathsf{c}\right)}{\widehat{\mathbb{P}}\left(X_2 = \mathsf{c}\right)}$$

$$= \frac{\widehat{\mathbb{P}}(\mathsf{ccc}) + \widehat{\mathbb{P}}(\mathsf{vcc})}{\widehat{\mathbb{P}}(\mathsf{ccc}) + \widehat{\mathbb{P}}(\mathsf{ccv}) + \widehat{\mathbb{P}}(\mathsf{vcc}) + \widehat{\mathbb{P}}(\mathsf{vcv})} = \frac{3k + 8k}{3k + 8k + 8k + 3k - 1}$$

$$= \frac{11k}{22k - 1} \approx \frac{1}{2}, \quad \text{for large } k.$$

Since $\frac{3}{11} \neq \frac{11k}{22k-1}$, this chain doesn't satisfy the definition (1.1), and hence we have shown that this process is not a Markov chain. $\quad\square$

APPLICATION 1.3. According to the Mendelian model of gene inheritance in humans, named after Gregor Johann Mendel (1822-1884), a specific genetic trait is determined by a pair of genes, that can be of three types: AA, Aa, or aa. During reproduction, an offspring inherits one gene of the pair from each parent, and genes are selected at random, independently of each other.

Suppose gene a is a mutant gene or a person possessing this gene is a carrier of a disease. Consider the genotype of the offspring in successive generations if the second parent always has genotype AA. This can be presented as a Markov chain with the state space $S = \{AA, Aa, aa\}$ and transition probability matrix

$$\mathbf{P} = \begin{array}{c} \\ AA \\ Aa \\ aa \end{array} \begin{array}{ccc} AA & Aa & aa \\ \left[\begin{array}{ccc} 1 & 0 & 0 \\ 0.5 & 0.5 & 0 \\ 0 & 1 & 0 \end{array}\right]. \end{array}$$

Indeed, if parents have genes (AA, AA), then their offspring are bound to have genes AA with probability 1. If parents have genes (Aa, AA), they are equally likely to spit into AA or Aa, and finally, if parents have genes (aa, AA), their offspring with certainty will have genes Aa. The diagram for this model is

```
#specifying the transition probability matrix
tm<- matrix(c(1, 0, 0, 0.5, 0.5, 0, 0, 1, 0), nrow=3, ncol=3,
byrow=TRUE)

#transposing the transition probability matrix
tm.tr<- t(tm)
```

```
#plotting the diagram for the Markov chain
library(diagram)
plotmat(tm.tr, pos=c(1,2), name=c("AA", "Aa", "aa"),
arr.length=0.3, arr.width=0.1, box.col="light blue",
box.lwd=1, box.prop=0.5, box.size=0.12, box.type="circle",
cex.txt=0.8, lwd=1, self.cex=0.6, self.shiftx=0.17,
self.shifty=-0.01)
```

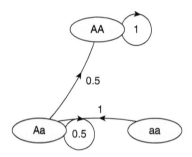

Note that states *Aa* and *aa* are transient states, and *AA* is the absorbing state. It means that in a long run, the gene *a* will disappear from the population. To convince ourselves, we compute the stationary distribution.

```
#creating Markov chain object
library(markovchain)
mc<- new("markovchain", transitionMatrix=tm,
states=c("AA", "Aa", "aa"))

#computing stationary distribution
steadyStates(mc)
```

AA Aa aa
 1 0 0

Let's assume that initially the gene *a* is present in the 1% of the population in the combination *aa*, that is, $p_{AA} = 0.99, p_{Aa} = 0$, and $p_{aa} = 0.01$. We run the R code below to see how the population genetic composition changes from generation to generation by computing recursively $p_{n+1} = p_n \cdot \mathbf{P}, n = 1, 2, 3, etc.$, with the initial condition $p_1 = (0.99, 0, 0.01)$.

```
library(expm)
gen1<- c(0.99, 0, 0.01)
gen<- gen1%*%tm
for (n in 2:10) {
print(n)
print(round(gen, digits=10))
gen<- gen%*%tm
}
```

n	p_{AA}	p_{Aa}	p_{aa}
1	0.99	0	0.01
2	0.99	0.01	0
3	0.995	0.005	0
4	0.9975	0.0025	0
5	0.99875	0.00125	0
6	0.999375	0.000625	0
7	0.9996875	0.0003125	0
8	0.9998438	0.00015625	0
9	0.999921875	0.000078125	0
10	0.9999609375	0.0000390625	0

Note that already in the second generation the gene type aa disappears, and is transformed into the hybrid type Aa, and that after as many as ten generations, the gene a is still lingering on in the population. □

APPLICATION 1.4. In this application, we present yet another chain that is not Markov. It has to do with weather conditions. On an intuitive level, weather tomorrow depends not just on today's weather but on the weather for several past days, if not the entire history of weather conditions in the region. To support this intuitive supposition numerically, we downloaded from *kaggle.com* an open-access historical hourly weather data for 2012-2017 (file "weather_description.csv"), focused only on the column for Los Angeles, and clumped the weather conditions into the four categories

$$S = \{s="\text{sky clear}", c="\text{cloudy}", f="\text{fog}", r="\text{rain}"\}.$$

We then found empirically the conditional probability of clear sky tomorrow, given clear skies yesterday and today,

$$\widehat{\mathbb{P}}(X_3 = s \mid X_2 = s, X_1 = s) = \frac{\widehat{\mathbb{P}}(sss)}{\widehat{\mathbb{P}}(sss) + \widehat{\mathbb{P}}(ssc) + \widehat{\mathbb{P}}(ssf) + \widehat{\mathbb{P}}(ssr)} = 0.9647,$$

and the conditional probability of clear sky tomorrow, given clear sky today,

$$\widehat{\mathbb{P}}(X_3 = s \mid X_2 = s) = \frac{\widehat{\mathbb{P}}(sss) + \widehat{\mathbb{P}}(css) + \widehat{\mathbb{P}}(fss) + \widehat{\mathbb{P}}(rss)}{\widehat{\mathbb{P}}(sss) + \widehat{\mathbb{P}}(css) + \cdots + \widehat{\mathbb{P}}(rsr)} = 0.9356.$$

Since these two estimates are not the same, we concluded that hourly weather conditions don't form a Markov chain. The complete R code and output follow.

```
weather.data<- read.csv("./weather_description.csv",
header=TRUE, sep=",")


LA<- weather.data$Los.Angeles


X3<- ifelse(LA=="sky is clear", "clear", ifelse(LA %in%
c("broken clouds", "few clouds", "overcast clouds",
"scattered clouds"), "clouds", ifelse(LA %in% c("light
intensity drizzle", "dust", "fog", "haze", "mist", "smoke",
"drizzle"), "fog", "rain")))


library(Hmisc) #Harrell Miscellaneous packages
X2<- Lag(X3,shift=1)
X1<- Lag(X3,shift=2)


sss<- ifelse(X1=="clear"& X2=="clear"& X3=="clear",1,0)
css<- ifelse(X1=="clouds"& X2=="clear"& X3=="clear",1,0)
fss<- ifelse(X1=="fog"& X2=="clear"& X3=="clear",1,0)
rss<- ifelse(X1=="rain"& X2=="clear"& X3=="clear",1,0)
ssc<- ifelse(X1=="clear"& X2=="clear"& X3=="cloudy",1,0)
csc<- ifelse(X1=="clouds"& X2=="clear"& X3=="clouds",1,0)
fsc<- ifelse(X1=="fog"& X2=="clear"& X3=="clouds",1,0)
rsc<- ifelse(X1=="rain"& X2=="clear"& X3=="clouds",1,0)
ssf<- ifelse(X1=="clear"& X2=="clear"& X3=="fog",1,0)
csf<- ifelse(X1=="clouds"& X2=="clear"& X3=="fog",1,0)
fsf<- ifelse(X1=="fog"& X2=="clear"& X3=="fog",1,0)
rsf<- ifelse(X1=="rain"& X2=="clear"& X3=="fog",1,0)
ssr<- ifelse(X1=="clear"& X2=="clear"& X3=="rain",1,0)
csr<- ifelse(X1=="clouds"& X2=="clear"& X3=="rain",1,0)
fsr<- ifelse(X1=="fog"& X2=="clear"& X3=="rain",1,0)
rsr<- ifelse(X1=="rain"& X2=="clear"& X3=="rain",1,0)


#computing P(X3=s|X2=s,X1=s)
sum(sss)/sum(sss+ssc+ssf+ssr)
```

0.9647471

```
#computing P(X3=s|X2=s)
sum(sss+css+fss+rss)/sum(sss+css+fss+rss+ssc+csc+fsc+rsc+ssf
+csf+fsf+rsf+ssr+csr+fsr+rsr)
```

0.9355981

□

Exercises

EXERCISE 1.1. A Markov chain has a one-step transition probability matrix

$$
\begin{array}{c c c c}
 & 1 & 2 & 3 \\
1 & \begin{bmatrix} 0.3 & 0.4 & 0.3 \\ 0.2 & 0.3 & 0.5 \\ 0.8 & 0.1 & 0.1 \end{bmatrix} \\
2 & & & \\
3 & & &
\end{array}.
$$

Compute the following probabilities:
(a) $\mathbb{P}(X_3 = 2 \mid X_0 = 1, X_1 = 2, X_2 = 3)$.
(b) $\mathbb{P}(X_4 = 3 \mid X_0 = 2, X_3 = 1)$.
(c) $\mathbb{P}(X_0 = 1, X_1 = 2, X_2 = 3, X_3 = 1)$. Assume $\mathbb{P}(X_0 = 1) = 1$.
(d) $\mathbb{P}(X_0 = 1, X_1 = 2, X_3 = 3, X_5 = 1)$. Assume $\mathbb{P}(X_0 = 1) = 1$.

EXERCISE 1.2. Consider a Markov chain with the transition probability matrix

$$
\begin{array}{c c c c c c}
 & 1 & 2 & 3 & 4 & 5 \\
1 & \begin{bmatrix} 1.0 & 0.0 & 0.0 & 0.0 & 0.0 \\ 0.5 & 0.0 & 0.0 & 0.0 & 0.5 \\ 0.2 & 0.0 & 0.0 & 0.0 & 0.8 \\ 0.0 & 0.0 & 1.0 & 0.0 & 0.0 \\ 0.0 & 0.0 & 0.0 & 1.0 & 0.0 \end{bmatrix} \\
2 & & & & & \\
3 & & & & & \\
4 & & & & & \\
5 & & & & &
\end{array}.
$$

(a) Plot a diagram of the Markov chain.
(b) Identify all transient and recurrent classes. Identify all absorbing and reflective states. Find the period of each state.
(c) Simulate three trajectories of the chain that start at a randomly chosen state. Comment on what you see.
(d) Find the steady-state probabilities and interpret them. Is it an ergodic chain?
(e) Plot the unconditional probabilities at time n against the time and comment on how fast the probabilities converge to the steady-state distribution.

EXERCISE 1.3. Consider a Markov chain with the one-step transition probability matrix

$$
\begin{array}{c}
\begin{array}{ccccccc} 1 & 2 & 3 & 4 & 5 & 6 & 7 \end{array} \\
\begin{array}{c} 1 \\ 2 \\ 3 \\ 4 \\ 5 \\ 6 \\ 7 \end{array}
\left[\begin{array}{ccccccc}
0 & 1 & 0 & 0 & 0 & 0 & 0 \\
1 & 0 & 0 & 0 & 0 & 0 & 0 \\
0 & 0 & 0 & 0.4 & 0.2 & 0.2 & 0.2 \\
0 & 0 & 0 & 0 & 0.2 & 0.4 & 0.4 \\
0.3 & 0 & 0 & 0.1 & 0.3 & 0.1 & 0.2 \\
0 & 0 & 0 & 0.2 & 0.2 & 0.3 & 0.3 \\
0 & 0 & 0 & 0.5 & 0.2 & 0.2 & 0.1
\end{array}\right].
\end{array}
$$

(a) Plot a diagram of the Markov chain.

(b) Identify all recurrent and transient classes. Find their periods. Are there any absorbing and reflecting states?

(c) Simulate two trajectories of the chain that start at a randomly selected state. Discuss what you see in the plot.

(d) Calculate the limiting probabilities and interpret them. Is the chain ergodic?

(e) Plot the unconditional probability vectors p_n against n and comment on the speed of convergence to the limiting distribution.

EXERCISE 1.4. Consider a Markov chain with the one-step transition probability matrix

$$
\begin{array}{c}
\begin{array}{ccccc} 1 & 2 & 3 & 4 & 5 \end{array} \\
\begin{array}{c} 1 \\ 2 \\ 3 \\ 4 \\ 5 \end{array}
\left[\begin{array}{ccccc}
0.1 & 0.2 & 0.3 & 0 & 0.4 \\
0 & 0.5 & 0.5 & 0 & 0 \\
0 & 1 & 0 & 0 & 0 \\
0 & 0 & 0 & 0 & 1 \\
0 & 0 & 0 & 0.6 & 0.4
\end{array}\right].
\end{array}
$$

(a) Plot the diagram of the Markov chain.

(b) Find all recurrent and transient classes and their periods. Are there any absorbing or reflecting states?

(c) Simulate several trajectories of the Markov chain and discuss the patterns that you see.

(d) Show that the chain is non-ergodic because there are two invariant probability measures. Which one of them is the stationary distribution?

(e) Plot the graphs of unconditional probabilities against time, assuming successively that the chain starts in states 1, 2, 3, 4, and 5. Interpret each graph.

EXERCISE 1.5. In a box there are two red (R), four blue (B), and eight green (G) balls. One ball is drawn at a time and its color is noted. Consider the stochastic process $\{X_n, \ n = 1, 2, ...\}$ with the state space $S = \{R, B, G\}$.

(a) Show that this process is <u>not</u> a Markov chain, if the drawing is done without replacement. Hint: Show, for instance, that $\mathbb{P}(X_3 = G \mid X_1 = R, X_2 = B) \neq \mathbb{P}(X_3 = G \mid X_1 = G, X_2 = B)$.
(b) Show that this process is a Markov chain, if the drawing is done with replacement.

EXERCISE 1.6. Consider a sequence of heads and tails obtained by a series of independent flips of a fair coin. Show that it can be modeled by a Markov chain with the state space $S = \{H, T\}$. Find the transition probability matrix and the limiting distribution.

EXERCISE 1.7. Assume that the usage of vowels and consonants in "Moby Dick" by Herman Melville can be modeled by a Markov chain.
(a) Find the transition probability matrix for Chapter 1 of this novel. Calculate the limiting probabilities and verify that they are equal to the overall proportions of vowels and consonants in the text.
(b) Do states in Chapter 2 follow the same transition probability matrix as those in Chapter 1?

EXERCISE 1.8. A student at a secretarial school typed the sentence "The quick brown fox jumped over the lazy dog" 500 times. Show that the resulting text cannot be modeled as a Markov chain.

EXERCISE 1.9. Consider the Mendelian gene inheritance model introduced in Application 1.3. Suppose that the second parent is chosen randomly from the gene pool with all types AA, Aa, or aa.
(a) Show that the genotype of the offspring follows a Markov chain with the state space $S = \{(AA, AA), (AA, Aa), (AA, aa), (Aa, AA), (Aa, Aa), (Aa, aa), (aa, AA), (aa, Aa), (aa, aa)\}$ and the transition probability matrix

	(AA, AA)	(AA, Aa)	(AA, aa)	(Aa, AA)	(Aa, Aa)	(Aa, aa)	(aa, AA)	(aa, Aa)	(aa, aa)
(AA, AA)	$1/3$	$1/3$	$1/3$	0	0	0	0	0	0
(AA, Aa)	$1/6$	$1/6$	$1/6$	$1/6$	$1/6$	$1/6$	0	0	0
(AA, aa)	0	0	0	$1/3$	$1/3$	$1/3$	0	0	0
(Aa, AA)	$1/6$	$1/6$	$1/6$	$1/6$	$1/6$	$1/6$	0	0	0
(Aa, Aa)	$1/12$	$1/12$	$1/12$	$1/6$	$1/6$	$1/6$	$1/12$	$1/12$	$1/12$
(Aa, aa)	0	0	0	$1/6$	$1/6$	$1/6$	$1/6$	$1/6$	$1/6$
(aa, AA)	0	0	0	$1/3$	$1/3$	$1/3$	0	0	0
(aa, Aa)	0	0	0	$1/6$	$1/6$	$1/6$	$1/6$	$1/6$	$1/6$
(aa, aa)	0	0	0	0	0	0	$1/3$	$1/3$	$1/3$

(b) Determine the transient and recurrent classes of the Markov chain.
(c) Find the stationary distribution. What is the steady-state genetic composition for both parents?

(d)   Which initial state achieves the stationary distribution in the smallest number of generations? Which in the largest? Assume the precision of four correct decimals after rounding.

EXERCISE 1.10. Refer to Application 1.4. Consider the data in the file "weather_description.csv." Choose a city other than Los Angeles and conduct the analysis similar to the one given the application.

EXERCISE 1.11. On an intuitive level, pollution level for a region doesn't follow a Markov chain as the pollution level tomorrow depends on pre-history and not just on today's level. But suppose that for some areas, air quality status (good/unhealthy/hazardous) depends only on those in the previous two days, and not in earlier days. Show that in this case, we can look at two days at a time and model data as a Markov chain.

EXERCISE 1.12. Consider a simplified monopoly game with only five squares and respective incomes of $200, $0, -$75, $105, and -$130. A player starts at the first square, rolls a fair die once, and moves forward as many steps as the die shows.
(a) Argue that this game can be modeled as a Markov chain and find its transition probability matrix.
(b) Compute the steady-state probability of each square, and find the long-run winning of the player.

EXERCISE 1.13. Suppose that road traffic conditions can be modeled as a Markov chain with the state space $S = \{\texttt{light}, \texttt{heavy}, \texttt{jammed}\}$, and suppose that traffic conditions change every 20 minutes. Assume that between 1PM and 4PM, the transition probability matrix is $\begin{bmatrix} 0.4 & 0.4 & 0.2 \\ 0.3 & 0.5 & 0.2 \\ 0 & 0.5 & 0.5 \end{bmatrix}$, whereas between 4PM and 6PM it changes to $\begin{bmatrix} 0.1 & 0.5 & 0.4 \\ 0.1 & 0.3 & 0.6 \\ 0 & 0.1 & 0.9 \end{bmatrix}$.
(a) If the traffic starts with the $\texttt{light}$ state at 1PM, what is the distribution of the states at 6PM?
(b) Simulate 10,000 trajectories to verify the result of part (a).

EXERCISE 1.14. A certain species of shrubs has four states: state 1 if it is sustainable, state 2 if it is threatened, state 3 if it is endangered, and state 4 if it is extinct. Plant assessment surveys are done at regular time intervals.

Transitions between states are modeled by a Markov chain with the transition probability matrix

$$
\begin{array}{cccc}
 & 1 & 2 & 3 & 4 \\
\begin{array}{c} 1 \\ 2 \\ 3 \\ 4 \end{array} &
\left[\begin{array}{cccc}
0.6 & 0.2 & 0.1 & 0.1 \\
0.7 & 0.2 & 0.1 & 0.0 \\
0.1 & 0.3 & 0.4 & 0.2 \\
0.0 & 0.0 & 0.0 & 1.0
\end{array}\right].
\end{array}
$$

(a) Assuming that a shrub is initially sustainable, simulate several trajectories of the Markov chain.

(b) Find the probability that initially sustainable shrub will eventually become extinct.

EXERCISE 1.15. A music instrument store is open every day of the week except Monday. During that day, if the inventory count is below 3, more instruments are ordered, so that by Tuesday morning there are 7 instruments in stock. If 3 or more instruments are in stock, then no action is taken. The number of instruments sold during the business days is a Poisson random variable with a mean of 4. Any demand that cannot be satisfied is lost. (a) Argue that the inventory each Tuesday morning can be modeled as a Markov chain. Find its state space and the one-step transition probability matrix.

(b) Generate inventory trajectories, assuming that the initial inventory size is randomly chosen.

(c) Suppose one week there are 7 instruments in stock on Tuesday morning. Compute the probability that there will be 7 instruments in stock also on each of the three subsequent Tuesday mornings.

(d) The weekly storage cost is $5 per instrument that is in the store on Tuesday morning. Compute the long-run expected weekly storage cost.

2

Random Walk

2.1 Definition of Random Walk

Consider a random process which state space comprises all integers on a real line. The process starts at zero and transitions either one step to the right with probability p or one step to the left with probability $1 - p$. This process is called a *simple random walk*. It is called *symmetric* if $p = 1/2$ and *asymmetric*, otherwise. It is also termed *infinite* because it is defined on an infinite set of integers. If a random walk occurs on a finite subset of integers, it is termed a *finite-state random walk* or a *random walk on a finite grid* or a *bordered random walk*.

In this chapter, we also consider some variations of a random walk, such as two-, three-, and higher-dimensional random walks, and random walks on graphs. Note that in general, random walks don't have to start at the origin. They can start at any randomly chosen starting point. Below we give formal definitions of all these processes.

A *simple (asymmetric, one-dimensional, infinite) random walk*[1] is a special case of a Markov chain which state space consists of integers $S = \{0, \pm1, \pm2, \ldots\}$, and the transition probabilities are of the form $p_{i,i+1} = \mathbb{P}(X_{n+1} = i + 1 \mid X_n = i) = p$ and $p_{i,i-1} = \mathbb{P}(X_{n+1} = i - 1 \mid X_n = i) = 1 - p$, $i = 0, \pm1, \pm2, \ldots$. The transition probability matrix for a random walk is

$$
\begin{array}{c}
 \\
\begin{array}{r}
-2 \\
-1 \\
0 \\
1 \\
2 \\
\end{array}
\begin{bmatrix}
\cdots & 0 & p & 0 & 0 & 0 & \cdots \\
\cdots & 1-p & 0 & p & 0 & 0 & \cdots \\
\cdots & 0 & 1-p & 0 & p & 0 & \cdots \\
\cdots & 0 & 0 & 1-p & 0 & p & \cdots \\
\cdots & 0 & 0 & 0 & 1-p & 0 & \cdots \\
\end{bmatrix}
\end{array}.
$$

A *symmetric random walk in two dimensions* is defined as a random walk on an integer lattice that moves right or left or up or down with a probability

---

[1] The term was first used in Pearson K. (1905). "The Problem of the Random Walk". *Nature*, 72: 294.

DOI: 10.1201/9781003244288-2

1/4. Likewise, a *symmetric random walk in d dimensions* is a random walk on the d-dimensional integer lattice, with equiprobable moves in $2d$ directions (with probability $1/(2d)$).

Some variations of a simple random walk include a *random walk with loops (or delays)*, in the sense that the allowed moves are to the right with probability p, to the left with probability q, and remaining in the same state with probability $1 - p - q$. Also, in two dimensions, it might be allowed to move diagonally as well as to stay in the same place, so each move has a probability of $1/9$. Or a *finite-state random walk* might be defined on a finite integer grid with the boundary (or border, or barrier) states being either reflecting or absorbing. In addition, it is possible to define a *random walk!on a graph*, where at every state, the process chooses among all neighboring states with equal probability. Below we consider some examples.

EXAMPLE 2.1. A gambler either wins \$5 with probability 0.55 or loses \$5 with probability 0.45. He starts playing with \$50 and plays until he either goes broke or doubles his original amount. This is an example of a finite one-dimensional random walk with absorbing barriers, since, once entered, the states \$0 and \$100 are never left. The transition probability matrix is

	\$0	\$5	\$10	...	\$45	\$50	\$55	...	\$95	\$100
\$0	1	0	0	...	0	0	0	...	0	0
\$5	0.45	0	0.55	...	0	0	0	...	0	0
\$10	0	0.45	0	...	0	0	0	...	0	0
...		...		...		...		...		...
\$45	0	0	0	...	0	0.55	0	...	0	0
\$50	0	0	0	...	0.45	0	0.55	...	0	0
\$55	0	0	0	...	0	0.45	0	...	0	0
...		...		...		...		...		...
\$95	0	0	0	...	0	0	0	...	0	0.55
\$100	0	0	0	...	0	0	0	...	0	1

If, for example, a gambler who goes broke can take a credit that brings him back to the \$50 fortune, the state \$0 is no longer an absorbing state, but is, in fact, a reflecting state. On the other end of his wealth spectrum, once the gambler reaches the \$100 fortune, he returns the credit (with, say, 10% interest) and is bounced to \$45, and thus the \$100 state is reflecting as well. □

EXAMPLE 2.2. Suppose that a mouse is running through a maze with rooms A, B, C, D, and E, and two exits (see the illustration by Shayan Khatri). The mouse runs along the passages, and in any given room, it decides randomly which passage to take.

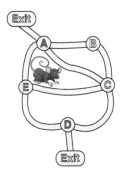

This motion can be modeled as a random walk on a graph. *Exit* is an absorb-ing state. State A has four neighboring states: *Exit*, B, C, and E, and thus the mouse can go to either one of them with a probability of $1/4$. State B has two neighboring states, A and C, and the mouse goes to either one with a probability of $1/2$. State C has connections to all the other four states, giving the mouse a probability of $1/4$ to choose the direction. State D connects to three states, C, E, and *Exit*, so the mouse selects the state to go to next with a probability of $1/3$. Finally, from state E the mouse can reach three states, A, C, or D, with probability $1/3$ each. The one-step transition matrix summarizes these probabilities.

$$
\begin{array}{c c}
 & \begin{array}{cccccc} Exit & A & B & C & D & E \end{array} \\
\begin{array}{c} Exit \\ A \\ B \\ C \\ D \\ E \end{array} &
\left[\begin{array}{cccccc}
1 & 0 & 0 & 0 & 0 & 0 \\
1/4 & 0 & 1/4 & 1/4 & 0 & 1/4 \\
0 & 1/2 & 0 & 1/2 & 0 & 0 \\
0 & 1/4 & 1/4 & 0 & 1/4 & 1/4 \\
1/3 & 0 & 0 & 1/3 & 0 & 1/3 \\
0 & 1/3 & 0 & 1/3 & 1/3 & 0
\end{array}\right].
\end{array}
$$

□

2.2 Must-Know Facts About Random Walk

Denote by $\{X_n,\ n = 0, 1, 2, \dots\}$ an asymmetric one-dimensional random walk that moves in the positive direction with probability p. Assume that the ran-dom walk starts at zero, that is, $X_0 = 0$. We formulate some interesting results and present them as propositions.

PROPOSITION 2.1. (MEAN AND VARIANCE OF A RANDOM WALK). The mean and variance of a random walk at time n are $\mathbb{E}(X_n) = (2p-1)n$ and $\mathbb{V}ar(X_n) = 4p(1-p)n$.

PROOF: We can write $X_n = Z_1 + \cdots + Z_n$ where Z_i's are independent random variables with the binary distribution:

$$Z_i = \begin{cases} 1, & \text{with probability } p, \\ -1, & \text{with probability } 1-p, \end{cases} \qquad i = 1, \ldots, n.$$

Note that $\mathbb{E}(Z_i) = (1)(p) + (-1)(1-p) = 2p-1$, and $\mathbb{V}ar(Z_n) = (1)^2(p) + (-1)^2(1-p) - (2p-1)^2 = 1 - (2p-1)^2 = 4p - 4p^2 = 4p(1-p)$. Thus, we compute $\mathbb{E}(X_n) = \mathbb{E}(Z_1) + \cdots + \mathbb{E}(Z_n) = (2p-1)n$, and $\mathbb{V}ar(X_n) = \mathbb{V}ar(Z_1) + \cdots + \mathbb{V}ar(Z_n) = 4p(1-p)n$. $\square$

Note that for a symmetric random walk, $p = 1/2$, and hence, $\mathbb{E}(X_n) = 0$ and $\mathbb{V}ar(X_n) = n$, and after n steps, a typical distance from the origin is on the order of $\sqrt{n}$.

PROPOSITION 2.2. (PROBABILITY OF A RETURN). The probability that a one-dimensional random walk returns to the starting point in exactly n steps is $\binom{n}{n/2} p^{n/2}(1-p)^{n/2}$ if n is even, and zero, otherwise.

PROOF: First of all, note that the random walk can return to the starting point only in an even number of steps, half of which it should be moving to the right (with probability p), and half, to the left (with probability $1-p$). There are $\binom{n}{n/2}$ paths with this property, hence:

$$\mathbb{P}(X_n = 0 \mid X_0 = 0) = \begin{cases} \binom{n}{n/2} p^{n/2}(1-p)^{n/2}, & \text{if } n \text{ is even,} \\ 0, & \text{if } n \text{ is odd} \end{cases}.$$

Note that the random walk can cross the starting point several times before ending there in exactly n steps. $\square$

The proofs of the next two propositions are omitted as they lie beyond the scope of this book.

PROPOSITION 2.3. (RECURRENCE VS. TRANSIENCE OF 1D RANDOM WALK). A one-dimensional random walk will come back to the origin infinitely many times with probability one if and only if $p = 1/2$. This means that a symmetric one-dimensional random walk is *recurrent*, while an asymmetric one-dimensional random walk is *transient*. It will, with a positive probability, come back only finite number of times and will eventually wander away.

PROPOSITION 2.4. (RECURRENCE VS. TRANSIENCE OF d-DIM RANDOM WALK). A two-dimensional random walk is recurrent if and only if it is symmetric (i.e., it goes up or down or left or right with probability $1/4$). Any random walk (symmetric or not) in 3D or a higher dimension is transient. It will eventually wander away from the origin with a positive probability.

Thence, a bug crawling randomly along a railroad track will visit every inch of it infinitely many times. Likewise, a King moving randomly on an infinite chessboard will visit every square infinitely many times, whereas, say, a drone performing a random walk in the sky will eventually go into outer space, never to be seen again.

2.3   Simulations in R

SIMULATION 2.1. Below we simulate three trajectories of a one-dimensional random walk. We assume the walk starts at zero, and specify p as 0.6 and the number of steps as 25.

```
#specifying parameters
ntraj<- 3
p<- 0.6
nsteps<- 25

#specifying seed
set.seed(45568223)

#defining walk as matrix
walk<- matrix(NA, nrow=nsteps, ncol=ntraj)

#simulating trajectories
for (j in 1:ntraj) {
walk[1,j]<- 0
for (i in 2:nsteps)
walk[i,j]<- ifelse(runif(1)<p, walk[i-1,j]+1, walk[i-1,j]-1)
}

#plotting trajectories
matplot(walk, type="l", lty=1, lwd=2, col=2:4,
ylim=c(range(walk)), xlab="Step", ylab="Position",
panel.first=grid())
```

```
points(1:nsteps, walk[,1], pch=16, col=2)
points(1:nsteps, walk[,2], pch=16, col=3)
points(1:nsteps, walk[,3], pch=16, col=4)
```

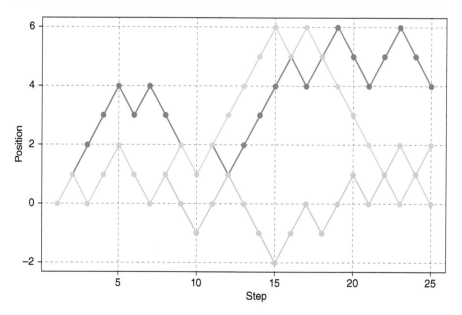

☐

SIMULATION 2.2. The code below simulates and plots a two-dimensional random walk with a total of 10,000 steps, emanating from the origin.

```
#specifying number of steps
nsteps<- 10000

#specifying seed
set.seed(607335)

#defining walk as matrix
walk<- matrix(NA, nrow=nsteps, ncol=2)

#setting starting point
walk[1,]<- c(0,0)

#definiting random steps
rstep<- matrix(c(1, 0, -1, 0, 0, 1, 0, -1), nrow=4, ncol=2,
byrow=TRUE)
```

```
#simulating trajectories
for (i in 2:nsteps)
walk[i,]<- walk[i-1,] + rstep[sample(1:4, size=1),]

#plotting trajectories
plot(x=walk[,1], y=walk[,2], type="l", col="blue",
xlim=range(walk[,1]), ylim=range(walk[,2]), xlab="x",
ylab="y", panel.first=grid())

#adding starting point
points(cbind(walk[1,1], walk[1,2]), pch=16, col="green",
cex=2)

#adding ending point
points(cbind(walk[nsteps,1],walk[nsteps,2]), pch=16,
col="red", cex=2)
```

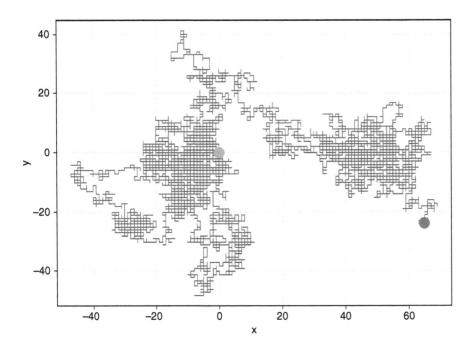

□

SIMULATION 2.3. The following code simulates and plots a three-dimensional random walk with 5,000 steps.

```
#specifying number of steps
nsteps<- 5000

#specifying seed
set.seed(830126)

#defining walk as matrix
walk<- matrix(NA, nrow=nsteps, ncol=3)

#setting starting point
walk[1,]<- c(0,0,0)

#defining random steps
rstep<- matrix(c(1,0,0,-1,0,0,0,1,0,0,-1,0,0,0,1,0,0,-1),
nrow=6, ncol=3, byrow=TRUE)

#simulating trajectories
for (i in 2:nsteps)
walk[i,]<- walk[i-1,]+rstep[sample(1:6, size=1),]

#plotting trajectories
library(plot3D)

lines3D(walk[,1], walk[,2], walk[,3], col="blue",
xlim=range(walk[,1]), ylim=range(walk[,2]),
zlim=range(walk[,3]), xlab="x", ylab="y", zlab="z", bty="b2",
ticktype="detailed")

#adding starting point
points3D(x=walk[1,1], y=walk[1,2], z=walk[1,3], add=TRUE,
pch=16, col="green", cex=2)

#adding ending point
points3D(walk[nsteps,1], walk[nsteps,2], walk[nsteps,3],
add=TRUE, pch=16, col="red", cex=2)
```

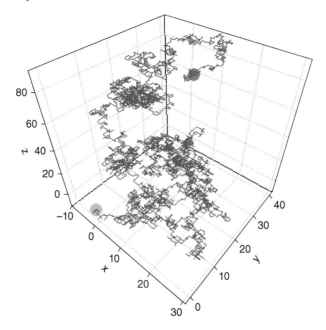

2.4 Applications of Random Walk

APPLICATION 2.1. (GAMBER'S RUIN PROBLEM). Gambling is perhaps the oldest area of application of random walks. And the most famous is the *Gambler's Ruin Problem*. A version of the gambler's ruin problem has been formulated as early as 1656, in correspondence between Blaise Pascal and Pierre de Fermat.

Suppose a gambler starts with a fortune of i and will move up \$1 with probability p or down \$1 with probability $q = 1 - p$ until he is either broke or reaches the fortune of N. What is the probability that he goes broke?

To answer this question, we will compute the complementary probability of reaching the fortune of N. Denote by $\mathbf{P}_j$ the probability of winning N if a gambler starts with j. Conditioning on the outcome of the first move, we can write the recurrence relation: $\mathbf{P}_j = q\mathbf{P}_{j-1} + p\mathbf{P}_{j+1}$ with the boundary conditions $\mathbf{P}_0 = 0$ and $\mathbf{P}_N = 1$. We can write this relation as $p\mathbf{P}_j + q\mathbf{P}_j = q\mathbf{P}_{j-1} + p\mathbf{P}_{j+1}$ and rewrite as $\mathbf{P}_{j+1} - \mathbf{P}_j = \frac{q}{p}\left(\mathbf{P}_j - \mathbf{P}_{j-1}\right)$. From here, we see that $\mathbf{P}_2 - \mathbf{P}_1 = \frac{q}{p}(\mathbf{P}_1 - \mathbf{P}_0) = \frac{q}{p}\mathbf{P}_1$, $\mathbf{P}_3 - \mathbf{P}_2 = \frac{q}{p}(\mathbf{P}_2 - \mathbf{P}_1) = \left(\frac{q}{p}\right)^2 \mathbf{P}_1, \ldots, \mathbf{P}_i - \mathbf{P}_{i-1} = \left(\frac{q}{p}\right)^{i-1} \mathbf{P}_1$. Summing up the identities, we obtain $\mathbf{P}_i - \mathbf{P}_1 = \left[\frac{q}{p} + \left(\frac{q}{p}\right)^2 + \cdots + \left(\frac{q}{p}\right)^{i-1}\right] \mathbf{P}_1$. From here,

$$\mathbf{P}_i = \begin{cases} \frac{1-(q/p)^i}{1-q/p}\,\mathbf{P}_1, & \text{if } q/p \neq 1, \\ i\mathbf{P}_1, & \text{if } q/p = 1. \end{cases}$$

To find $\mathbf{P}_1$, we use the boundary condition $\mathbf{P}_N = 1$. It yields

$$\mathbf{P}_1 = \begin{cases} \frac{1-q/p}{1-(q/p)^N}, & \text{if } q/p \neq 1, \\ 1/N, & \text{if } q/p = 1. \end{cases}$$

Consequently, if gambling is modeled as a symmetric random walk (with $p = q = 1/2$), the probability of reaching the fortune \$$N$ is i/N, and thus, the probability of ruin is $(N-i)/N$. If the model is an asymmetric random walk, the probability of reaching \$$N$ is $\mathbf{P}_i = \dfrac{1-(q/p)^i}{1-(q/p)^N}$, and hence, the probability of ruin is $1 - \mathbf{P}_i = \dfrac{(q/p)^i - (q/p)^N}{1-(q/p)^N}$.

Also, it can be shown (see Exercise 2.7) that the expected number of games that the gambler plays until he reaches \$$N$ or goes bankrupt is

$$\mathbf{E}_i = \begin{cases} \frac{N}{q-p}\frac{(q/p)^i-(q/p)^N}{1-(q/p)^N} - \frac{N-i}{q-p} = \frac{i-N\mathbf{P}_i}{q-p}, & \text{if } q/p \neq 1, \\ i(N-i), & \text{if } q/p = 1. \end{cases}$$

Let us see how it plays out with some specific values. In Example 2.1, the gambler started with \$50, goes up \$5 with probability 0.55, or goes down \$5 with probability 0.45. He would end up with either \$100 or \$0. Since the increment is \$5, in our notation this translates into $i = 10$, $N = 20$, and $p = 0.55$. The probability of reaching \$100 is $\dfrac{1-(0.45/0.55)^{10}}{1-(0.45/0.55)^{20}} = 0.8815$ and the probability of ruin is, respectively, $1 - 0.8815 = 0.1185$.

As for the expected number of games, we compute $\mathbf{E}_i = \dfrac{10-(20)(0.8815)}{0.45-0.55} = 76.3$. That is, the gambler plays, on average, 76.3 games.

We can verify these probabilities and the expected number of games empirically, by running the following R code that simulates 100,000 trajectories and counts how many of them ended in 20, how many ended in 0, and keeps track of the total number of games played until the end is reached. These values are then averaged over the total number of trajectories.

```
#specifying parameters
p<- 0.55
i<- 10
N<- 20
ntraj<- 100000

#defining walk as vector
walk<- c()

#setting counters
nNs<- 0
nzeros<- 0
ngames<- 0

#setting seed number
set.seed(30112443)

#simulating trajectories until hitting N or 0
for (j in 1:ntraj) {
walk[1]<- i
k<- 2
repeat {
walk[k]<- ifelse(runif(1)<p, walk[k-1]+1, walk[k-1]-1)
ngames<- ngames + 1

    if (walk[k]==N) {
      nNs<- nNs+1
        break
        }

    else if(walk[k]==0) {
     nzeros<- nzeros+1
      break
        }

k<- k+1
}
}

print(prob.Ns<- nNs/ntraj)
```

0.88279

```
print(prob.zeros <- nzeros/ntraj)
```

0.11721

```
print(mean.ngames<- ngames/ntraj)
```

76.18866

Next, we can plot the graph of the probabilities as a function of p, for our specific values of $i = 10$ and $N = 20$. The syntax and the graph follow. The green curve depicts the probability of reaching N, whereas the red one displays the probability of ruin.

```
p<- seq(0.35,0.65,0.001)
i<- 10
N<- 20
q<- 1-p
p.ruin<- ifelse(p==0.5,(N-i)/N,((q/p)^i-(q/p)^N)/(1-(q/p)
^N))

#ploting the graphs
plot(p, p.ruin, type = "l", lwd=2, col = "red", xlab="p",
ylab="Probability", panel.first = grid())

lines(p, 1-p.ruin, lwd=2, col = "green")

legend("right", c("Probability of $0", "Probability of $N"),
lty=1, col=2:3)
```

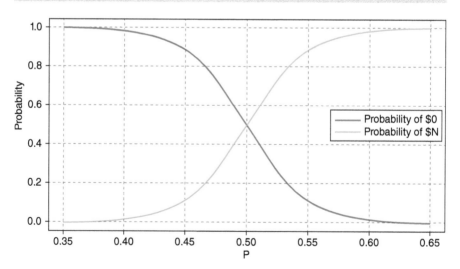

Note from the graph that the probability p doesn't have to be very small for the ruin to happen almost certainly. For p ranging between 0.4 and 0.6, the probability of ruin goes down from almost 1 to almost 0. Respectively, the probability of reaching N increases from almost 0 to almost 1.

Finally, we plot the graph of the expected number of games against p. The code and the output are presented below.

```
p<- seq(0.01,1,0.01)
i<- 10
N<- 20
q<- 1-p
E.ngames<-
ifelse(p==0.5,i*(N-i),(i-N*(1-(q/p)^i)/(1-(q/p) ^N))/(1-2*p))

plot(p, E.ngames, type="l", lwd=2, col="green", xlab="p",
ylab="Expected number of games", panel.first=grid())
```

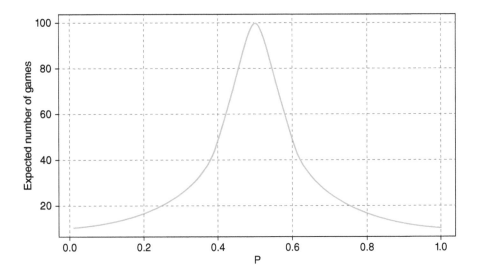

Note that the maximum of this function is achieved at $p = 1/2$ and the maximum value is $i(N - i) = (10)(20 - 10) = 100$. Also, the graph is symmetric due to the starting point being right in the middle, i.e., $i = N/2$. □

APPLICATION 2.2. (RANDOM WALK ON A GRAPH). Consider the random walk through the maze introduced in Example 2.2. Suppose the mouse starts in room C and spends one second on each transition from room to room (or *Exit*). We want to calculate how many seconds, on average, the mouse spends in the maze before exiting. We argue as follows. The mouse will spend exactly k seconds in the maze if it exits in exactly k transitions. So, we need to find the probability to transition from room C to an *Exit* in exactly k steps. Let $\mathbf{P}$ denote the one-step transition probability matrix. Then

$$(0,0,0,1,0,0)\,\mathbf{P}^k\,(1,0,0,0,0,0)^{-1}$$

gives the probability to transition from vertex C to *Exit* in k or fewer steps, and thus,

$$(0,0,0,1,0,0)\left(\mathbf{P}^k - \mathbf{P}^{k-1}\right)(1,0,0,0,0,0)^{-1}$$

is the probability to transition from C to *Exit* in exactly k steps. Consequently, the formula for the expected time that the mouse spends in the maze before it reaches an *Exit* is

$$\mathbb{E}(time\ to\ exit) = (0,0,0,1,0,0)\left((1)\mathbf{P} + (2)(\mathbf{P}^2 - \mathbf{P})\right.$$

$$\left. +(3)(\mathbf{P}^3 - \mathbf{P}^2) + (4)(\mathbf{P}^4 - \mathbf{P}^3) + \ldots\right)(1,0,0,0,0,0)^{-1}.$$

We run an R code to estimate this expected value. Convergence to the six decimal places that R outputs is achieved with the first 172 terms. Adding more terms doesn't change the output.

```
#specifying transition probability matrix
tm<- matrix(c(1,0,0,0,0,0,1/4,0,1/4,1/4,0,1/4,0,1/2,0,
1/2,0,0,0,1/4,1/4,0,1/4,1/4,1/3,0,0,1/3,0,1/3,0,1/3,0,1/3,1/
3,0), nrow=6, ncol=6, byrow=TRUE)

#setting counter
nsec<- 0

#estimating expected number of seconds
p<- matrix(NA, nrow=172, ncol=6)
p[1,]<- c(0, 0, 0, 1, 0, 0)

for (i in 2:172) {
   p[i,]<- p[i-1,]%*%tm
     nsec<- nsec+(i-1)*(p[i,1]-p[i-1,1])
     }

print(nsec)
```

9.967213

Hence, the mouse spends, on average, 9.967213 seconds in the maze, making that many transitions between the states (and an *Exit*). □

Exercises

EXERCISE 2.1. Consider a one-dimensional random walk with the transition probability $p = 0.3$. Simulate 10,000 trajectories of length 50 steps and calculate empirical mean and variance. Are the estimates close to the theoretical values?

EXERCISE 2.2. Consider a symmetric one-dimensional random walk that originates at 0.
(a) Simulate 10,000 trajectories with 1,000 steps each. How many of the trajectories are at point 0 on the 1,000th step?
(b) Find the theoretical probability of returning to 0 on the 1,000th step. Compare to the empirical value.

EXERCISE 2.3. Simulate 10,000 trajectories of 1D, 2D, and 3D symmetric random walks that start at the origin and continue for at most 1,000 steps.
(a) Compute how many of them returned to the origin at least once. Compare the results for different dimensions. Hint: Terminate a trajectory when it returns to the origin.
(b) Consider only the trajectories that returned to the origin within the 1,000 steps. Compute the average number of steps it took those trajectories to return to the origin. Compare the results for different dimensions.

EXERCISE 2.4. Simulate 10,000 trajectories of a two-dimensional symmetric random walk that starts at the origin and continues for a maximum of 1,000 steps.
(a) Estimate the probability of a trajectory ever hitting the vertical barrier $x = 30$.
(b) Estimate the average number of steps it takes a trajectory to hit the barrier, provided it did hit the barrier within the 1,000 steps.
(c) Estimate the expected value of the y-coordinate at the time when the random walk hits the barrier. What should this value be from the theoretical point of view? Hint: deduce from a symmetry argument.

EXERCISE 2.5. Simulate 100 trajectories of a two-dimensional symmetric random walk that starts at the origin and continues for 1,000 steps or until it hits a barrier. The value of the barrier varies between $x = 1$ and $x = 50$. Plot the empirical probability of hitting the barrier against the barrier value. Discuss the pattern you see.

EXERCISE 2.6. Simulate 1,000 trajectories of a two-dimensional symmetric random walk that starts at the origin and continues until it hits a side of a square centered at the origin and having a side length of 20. Estimate the average number of steps that it takes the random walk to reach the square.

EXERCISE 2.7. Suppose a gambler starts with a fortune of $\$i$ and will move up \$1 with probability p or down \$1 with probability $q = 1 - p$ until he either reaches the fortune of $\$B$ or is down to $\$A$.
(a)   Prove that the probability that he reaches $\$B$ before $\$A$ is

$$\mathbf{P}_i = \begin{cases} \frac{(q/p)^A - (q/p)^i}{(q/p)^A - (q/p)^B}, & \text{if } q/p \neq 1, \\ \frac{i-A}{B-A}, & \text{if } q/p = 1. \end{cases}$$

Hint: Show that $\mathbf{P}_i$ solves the recurrence relation $\mathbf{P}_i = p\mathbf{P}_{i+1} + q\mathbf{P}_{i-1}$, with the border constraints $\mathbf{P}_A = 0$ and $\mathbf{P}_B = 1$. Look for the solution in the form $\mathbf{P}_i = c(q/p)^i + d$, if $q/p \neq 1$, and $\mathbf{P}_i = ci + d$ if $q/p = 1$.

(b) Show that the expected number of games the gambler plays until he reaches $\$B$ or $\$A$ is

$$\mathbf{E}_i = \begin{cases} \frac{B-A}{q-p} \frac{(q/p)^i - (q/p)^B}{(q/p)^A - (q/p)^B} - \frac{B-i}{q-p}, & \text{if } q/p \neq 1, \\ (B-i)(i-A), & \text{if } q/p = 1. \end{cases}$$

Hint: Show that $\mathbf{E}_i$ satisfies the recurrence relation $\mathbf{E}_i = p\mathbf{E}_{i+1} + q\mathbf{E}_{i-1} + 1$ with the boundary conditions $\mathbf{E}_A = \mathbf{E}_B = 0$. Look for solutions in the form $\mathbf{E}_i = c(q/p)^i + d + i/(q-p)$, if $q/p \neq 1$, and $\mathbf{E}_i = ci + d - i^2$, if $q/p = 1$.

(c) Suppose a gambler comes to a casino with \$40 and plays a rigged game with $p = 0.47$ until he doubles the amount or is down to \$10 (to pay for a taxi). Calculate the probability that he walks out of the casino with \$80. How many games, on average, will he play? Simulate 10,000 trajectories and estimate the probability and the expected length of play.

EXERCISE 2.8. A student visits an Ancient History museum that is open between 9AM and 6PM. He enters the museum at 9AM and wanders the rooms in a random-walk fashion, spending 30 minutes in each room, and then choosing a door at random. The museum floor plan is given in the picture. How long will the student spend in the museum, on average? Does he expect to leave the museum before it closes for the day? Write down the formula and use R to calculate the result.

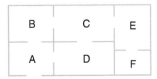

3

Poisson Process

3.1 Definition and Must-Know Facts About Poisson Process

A stochastic process $\{N(t),\, t \geq 0\}$ is called a *counting process* if $N(t)$ gives the total number of events occurring by time t.

A counting process $\{N(t),\, t \geq 0\}$ is said to have *independent increments* if the number of events that occur in non-overlapping time intervals are independent. For example, $N(5) - N(0)$, the number of events occurring between times 0 and 5, is independent of $N(10) - N(5)$, the number of events occurring between times 5 and 10.

A counting process $\{N(t),\, t \geq 0\}$ is said to have *stationary increments* if the distribution of the number of events that occur in any time interval depends only on the length of the interval. In other words, $N(t) - N(0)$ and $N(t+s) - N(s)$ have the same distribution that depends only on t and not on s.

A counting process $\{N(t),\, t \geq 0\}$ is called a *Poisson process*[1] with rate λ, if: (i) no events occur at time 0, i.e., $N(0) = 0$, (ii) it has independent increments, (iii) it has stationary increments, and (iv) $\mathbb{P}(N(t) = n) = \frac{(\lambda t)^n}{n!} e^{-\lambda t}$, $n = 0, 1, 2, \ldots$. Note that $\mathbb{E}(N(t)) = \mathbb{V}ar(N(t)) = \lambda t$.

Since the rate λ of a Poisson process is a constant not depending on time t, the process is sometimes referred to as a *homogeneous* (or *stationary*) Poisson process.

An *interarrival time* is the time between two consecutive occurrences of events. The interarrival time between $(n-1)$st and nth occurrences will be denoted by T_n, $n = 2, 3, \ldots$. The time of the first occurrence will be denoted by T_1.

---

[1] The reference to a Poisson process first appeared in two independent publications in 1940. The first was the article by Feller, W. (1940). "On the integro-differential equations of purely discontinuous Markov processes." *Trans. Am. Math. Society*, 48(3): 488 – 515. The second was the Ph.D. dissertation by Lundberg, O. (1940). "On random processes and their application to sickness and accident statistics." Uppsala: Almqvist & Wiksell.

DOI: 10.1201/9781003244288-3

The *waiting time* until the nth event (or *event time*), $S_n = T_1 + T_2 + \cdots + T_n$, is the time when the nth event occurs.

PROPOSITION 3.1. Interarrival times T_n, $n = 1, 2, \ldots$, are independent exponentially distributed random variables with the density function $f_{T_n}(t) = \lambda e^{-\lambda t}$, $t \geq 0$.

PROOF: Note that $\mathbb{P}(T_1 > t) = \mathbb{P}(N(t) = 0) = e^{-\lambda t}$. Therefore, $T_1 \sim Exp(\lambda)$. Next, conditioning on the value of the first occurrence, and using independence and stationarity of increments, we write

$$\mathbb{P}(T_2 > t) = \int_0^\infty \mathbb{P}(T_2 > t \mid T_1 = s) \, \lambda e^{-\lambda s} \, ds$$

$$= \int_0^\infty \mathbb{P}(N(t+s) - N(s) = 0 \mid N(s) = 1) \, \lambda e^{-\lambda s} \, ds$$

$$= \int_0^\infty \mathbb{P}(N(t) = 0) \, \lambda e^{-\lambda s} \, ds = e^{-\lambda t} \int_0^\infty \lambda e^{-\lambda s} \, ds = e^{-\lambda t},$$

that is, $T_2 \sim Exp(\lambda)$. More generally, conditioning on the time of the nth event occurrence and again using independence and stationarity of increments, we obtain

$$\mathbb{P}(T_{n+1} > t) = \int_0^\infty \mathbb{P}(T_{n+1} > t \mid S_n = s) \, f_{S_n}(s) \, ds$$

$$= \int_0^\infty \mathbb{P}(N(t+s) - N(s) = 0 \mid N(s) = n) \, f_{S_n}(s) \, ds$$

$$= \int_0^\infty \mathbb{P}(N(t) = 0) \, f_{S_n}(s) \, ds = e^{-\lambda t} \int_0^\infty f_{S_n}(s) \, ds = e^{-\lambda t}. \quad \square$$

REMARK 3.1. Recall that exponential is the only continuous distribution that possesses the *memoryless property*, $\mathbb{P}(T > t + s \mid T > t) = \mathbb{P}(T > s)$. In a Poisson process, the increments are independent and stationary and that implies that the process renews itself every moment. Therefore, on an intuitive level, the interarrival times should possess the memoryless property and thus be distributed exponentially. Also, if, say, on average, there are two occurrences per hour ($\lambda = 2$ per hour), the average waiting time between two occurrences is half an hour (mean$= 1/\lambda = 1/2$ hour).

PROPOSITION 3.2. The waiting time S_n has $Gamma(n, \lambda)$ distribution. Thus, $\mathbb{E}(S_n) = n/\lambda$ and $\mathbb{V}ar(S_n) = n/\lambda^2$. The density function is of the form $f_{S_n}(s) = \frac{\lambda^n s^{n-1}}{(n-1)!} e^{-\lambda s}$, $s \geq 0$.

PROOF: We can write S_n as the sum of interarrival times, i.e., $S_n = T_1 + T_2 + \cdots + T_n$. It is a general fact in the theory of probability that the sum of n independent exponentially distributed random variables with parameter λ has a gamma distribution with parameters n and λ. The quickest proof of this fact is through the moment generating functions. □

REMARK 3.2. Let's look at a Poisson process as a special case of a Markov chain. The state space of a Poisson process is $S = \{0, 1, 2, \dots\}$. The process jumps from initial state 0 to state 1 with probability 1, then to state 2 with probability 1, etc. The jumps are always of size 1, and transitions between states are allowed only in the direction of increase. Moreover, a Poisson process includes the time component, so it matters how long the process halts between jumps. In fact, we know that it halts an exponentially distributed time. This type of Markov chain is called a *continuous-time Markov chain*.

REMARK 3.3. It is essential to understand that Poisson arrivals occur one at a time. The probability of two simultaneous arrivals is zero since the interarrival time is exponentially distributed and thus the probability of an interarrival time being equal to zero is zero. It means that in situations when, say, people can potentially arrive in groups, one needs to count not arrivals of individual people, but count the arrival of a group as a single event. For example, if we model the process of people joining a ticket line in a movie theater, we would be likely to see an entire party joining the line, so we would count the party as one arrival.

EXAMPLE 3.1. A Poisson process is used to model occurrences of rare events. Here are some instances of Poisson processes: the number of people who enter a store or a bank or a restaurant or a gym or a National Park, the number of cars that pass a certain intersection, the number of auto accidents on a certain stretch of a freeway, the number of births in a hospital, the number of meteors in the night sky, or the number of phone calls to a credit card customer service. Typically, natural disasters occur according to a Poisson process: earthquakes, volcano eruptions, wildfires, etc. Of course, in all the above examples, the considered time period should be short enough for the rate of occurrence to be constant.

Some examples of processes that clearly are not governed by a Poisson law are events that happen according to a schedule, for example, the arrival of buses along a certain route, road closures due to construction work, quarry blasts, building demolitions. Also, events that happen in a competing market, where two rival companies might be scheduling two events at the same time. For instance, two pharmaceutical companies might simultaneously bring to the market two cardiac medications, or two production companies might release two movies on the same day. In addition, some periodic (or seasonal) natural phenomena cannot be modeled as a Poisson process, i.e., eruptions of Old Faithful geyser in Yellowstone Stone National Park, ocean tides, the appearance of sunspots, etc.   □

EXAMPLE 3.2. Tour buses arrive at a roadside mall with restaurants, bringing 50 tourists each. The times that elapse between consecutive arrivals are independent and exponentially distributed with mean of 15 minutes.

(a) On average, buses arrive every 15 minutes, so, for instance, the expected waiting time for the fifth bus to arrive is $(15)(5) = 75$ minutes, or 1 hour and 15 minutes. Now we put it in our theoretical framework. The rate of arrival of the buses is $\lambda = 4$ per hour. Denote by S_5 the time until the 5th bus arrives. We know that S_5 has a gamma distribution with parameters $n = 5$ and $\lambda = 4$. Therefore, $\mathbb{E}(S_5) = 5/4 = 1.25$ hours, or 1 hour and 15 minutes. We can also compute the variance of S_5 as $\mathbb{V}ar(S_5) = 5/4^2 = 0.3125$ hours squared, and the standard deviation as $\sqrt{\mathbb{V}ar(S_5)} = \sqrt{0.3125} = 0.559$ hours.

(b) The total expected number of tourists who are served lunch at the restaurants between, say, 11AM and 1:30PM is found as follows. Within the two and a half hours, we expect 10 bus arrivals, each carrying 50 tourists. Therefore, we expect a total of $(10)(50) = 500$ tourists. To write it formally, let $N(t)$ denote the number of buses that arrive by time t. We know that $N(t) \sim Poisson(4t)$. We are given that $t = 2.5$ hours, and therefore, the expected total number of tourists is $(50)\mathbb{E}(N(2.5)) = (50)(4)(2.5) = 500$.   □

EXAMPLE 3.3. Independence and stationarity of increments in a Poisson process allows elegant computations of conditional probabilities. For example, consider a Poisson process with rate $\lambda = 2.2$.

(a) Suppose we want to find the conditional probability that there will be 8 arrivals by time 5 given that there was 1 arrival by time 2. We can argue that since at time 2 the process renews itself, we need to compute the probability that within the next 3 time periods there will be 7 more arrivals. We write $\mathbb{P}(N(5) = 8 \,|\, N(2) = 1) = \mathbb{P}(N(5) - N(2) = 7 \,|\, N(2) = 1) = \{independence\} = \mathbb{P}(N(5) - N(2) = 7) = \{stationarity\} = \mathbb{P}(N(5 - 2) = 7) = \mathbb{P}(N(3) = 7) = \frac{((2.2)(3))^7}{7!} e^{-(2.2)(3)} = 0.147243$.

(b) Similarly, conditional expectations can be computed if we utilize independence and stationarity of increments. $\mathbb{E}[N(9) \,|\, N(7) = 10] = \mathbb{E}[N(9) - N(7) \,|\, N(7) = 10] + \mathbb{E}[N(7) \,|\, N(7) = 10] = \mathbb{E}[N(9) - N(7)] + 10 = \mathbb{E}[N(2)] + 10 = (2.2)(2) + 10 = 14.4$.

(c) Let S_{30} denote the time of occurrence of the 30th event. Independence and stationarity of increments helps us again to compute, for instance, the conditional expectation of S_{30} given that 8 events occurred in the first 12 time periods. We write $\mathbb{E}[S_{30} \mid N(12) = 8] = 12 + \mathbb{E}[time\ until\ 22\ more\ events] = 12 + \mathbb{E}[S_{22}] = 12 + 22/2.2 = 12 + 10 = 22.$ □

PROPOSITION 3.3. Suppose that in a Poisson process $\{N(t), t \geq 0\}$ with rate λ, an event can be either of category 1 with probability p, or of category 2 with probability $1 - p$. Denote by $N_1(t)$ and $N_2(t)$ the number of events of category 1 and 2 that occur by time t, respectively. Note that $N(t) = N_1(t) + N_2(t)$. Then, $\{N_1(t), t \geq 0\}$ and $\{N_2(t), t \geq 0\}$ are independent Poisson processes with respective rates λp and $\lambda (1 - p)$.

PROOF: Note that by definition, for an observed value of $N(t)$, $N_1(t)$ has a binomial distribution with parameters $N(t)$ and p. Respectively, $N_2(t)$ has a binomial distribution with parameters $N(t)$ and $1 - p$. Therefore, the joint probability distribution of $N_1(t)$ and $N_2(t)$ can be derived as

$$\mathbb{P}(N_1(t) = n_1,\ N_2(t) = n_2)$$

$$= \mathbb{P}(N_1(t) = n_1,\ N_2(t) = n_2 \mid N(t) = n_1 + n_2)\,\mathbb{P}(N(t) = n_1 + n_2)$$

$$= \binom{n_1 + n_2}{n_1} p^{n_1} (1 - p)^{n_2} \frac{(\lambda t)^{n_1 + n_2}}{(n_1 + n_2)!} e^{-\lambda t}$$

$$= \frac{(\lambda p t)^{n_1}}{n_1!} e^{-\lambda p t} \frac{(\lambda(1 - p)t)^{n_2}}{n_2!} e^{-\lambda(1-p)t}.$$

Hence, $N_1(t)$ and $N_2(t)$ are independent Poisson random variables with rates λp and $\lambda(1-p)$, respectively. Independence and stationarity of increments are inherited from those of the process $\{N(t), t \geq 0\}$. □

REMARK. In the above proposition, the Poisson process $\{N(t), t \geq 0\}$ is called the *superposition* of Poisson processes $\{N_1(t), t \geq 0\}$ and $\{N_2(t), t \geq 0\}$. In turn, $\{N_1(t), t \geq 0\}$ and $\{N_2(t), t \geq 0\}$ are called *thinned* (or *splitted*) Poisson processes.

EXAMPLE 3.4. Suppose phone calls to a customer service department in a credit card company arrive as a Poisson process with rate 3 per minute. Thirty percent of the calling customers experience technical difficulties when using their credit cards.

(a) The probability that during the next 15 minutes there will be 12 phone calls from customers who experience technical difficulties is computed as follows. We focus only on the customers who experience technical difficulties. Call this process $\{N_1(t),\ t \geq 0\}$. We know that it is a Poisson process with the rate $\lambda p = (3)(0.30) = 0.9$ per minute. So, we can write $\mathbb{P}(N_1(15) = 12) = \frac{((0.9)(15))^{12}}{12!}\, e^{-(0.9)(15)} = 0.10488$.

(b) Suppose now we want to calculate the probability that during the next 15 minutes there will be 40 phone calls from customers, half of whom experience technical difficulties. We denote by $\{N_2(t),\ t \geq 0\}$ the Poisson process that counts only the callers who don't experience technical difficulties. Its rate is $\lambda(1 - p) = (3)(0.7) = 2.1$. We know that the processes $\{N_1(t),\ t \geq 0\}$ and $\{N_2(t),\ t \geq 0\}$ behave independently. Hence, we write $\mathbb{P}(N_1(15) = 20,\ N_2(15) = 20) = \mathbb{P}(N_1(15) = 20)\,\mathbb{P}(N_2(15) = 20) = \frac{((0.9)(15))^{20}}{20!}\, e^{-(0.9)(15)} \cdot \frac{((2.1)(15))^{20}}{20!}\, e^{-(2.1)(15)} = (0.0228)(0.00794) = 0.000181.$ □

3.2   Simulations in R

We will present two simulation methods of a trajectory of a Poisson process.

SIMULATION 3.1. (EXPONENTIAL INTERARRIVALS). When simulating a trajectory of a Poisson process, we need first to simulate exponentially distributed interarrival times. We base our simulations on standard uniform random variables and use the inversion of the cumulative distribution function method to obtain exponentially distributed random variables: if $u \sim Unif(0, 1)$, then $-\frac{1}{\lambda}\ln(1 - u)$ is exponential with mean $1/\lambda$. Note that since $1 - u$ is also $Unif(0, 1)$, we can simplify the expression for the exponential random variables to $-\frac{1}{\lambda}\ln u$.

To plot the simulated trajectory, we use the `segment()` function which takes as arguments vectors of left and right endpoints. Below we present the code and graph of a trajectory of a Poisson process with rate 2 that stops when it makes the 20th jump.

```
#specifying parameters
lambda<- 2
njumps<- 20

#defining states
N<- 0:njumps
```

```
#setting time as vector
time<- c()

#setting initial value for time
time[1]<- 0

#specifying seed
set.seed(333422)

#simulating trajectory
for (i in 2:(njumps+1))
time[i]<- time[i-1]+round((-1/lambda)*log(runif(1)),2)

#plotting trajectory
# type="n" draws empty frame with no graph
plot(time, N, type="n", xlab="Time", ylab="State",
panel.first = grid())

segments(time[-length(time)], N[-length(time)],
time[-1]-0.07, N[-length(time)], lwd=2, col="blue")

points(time, N, pch=20, col="blue")
points(time[-1], N[-length(time)], pch=1, col="blue")
```

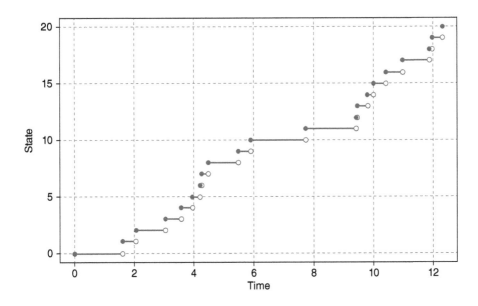

SIMULATION 3.2. (UNIFORM ORDER STATISTICS). First we need to prove the theoretical result. Consider a Poisson process $\{N(t),\ t \geq 0\}$. Given that n events occurred by time t, the times at which the events occurred are distributed as the order statistics of a uniform distribution on $(0, t)$.

To show this result, we derive the conditional density of n waiting times $S_1, S_2, \ldots, S_n$, using independence and exponential distribution of the inter-arrival times $T_1, T_2, \ldots, T_n$. We write

$$f_{S_1, S_2, \ldots, S_n}\left(s_1, s_2, \ldots, s_n \mid N(t) = n\right)$$

$$= \frac{f_{T_1}(s_1) f_{T_2}(s_2 - s_1) \cdots f_{T_n}(s_n - s_{n-1}) \mathbb{P}(T_{n+1} > t - s_n)}{\mathbb{P}(N(t) = n)}$$

$$= \frac{\lambda e^{-\lambda s_1} \lambda e^{-\lambda(s_2 - s_1)} \cdots \lambda e^{-\lambda(s_n - s_{n-1})} e^{-\lambda(t - s_n)}}{\frac{(\lambda t)^n}{n!} e^{-\lambda t}} = \frac{n!}{t^n}.$$

This gives us the following algorithm to generate trajectories:

Step 1. Fix t and generate $N(t) \sim Poi(\lambda t)$.
Step 2. Generate $N(t)$ standard uniform random variables $U_1, \ldots, U_{N(t)}$.
Step 3. Order $U_1, \ldots, U_{N(t)}$ in increasing order, obtaining the ordered set $U_{(1)}, \ldots, U_{(N(t))}$.
Step 4. Multiply the order statistics by t to obtain the set of event times $S_1 = t U_{(1)}, \ldots, S_{N(t)} = t U_{(N(t))}$.
Step 5. Define the states of the Poisson process as $N(0) = 0, N(S_1) = 1, N(S_2) = 2, \ldots, N(S_{N(t)}) = N(t)$.
Step 6. Plot the states against time.

The sample code and plot follow.

```
#specifying parameters
t<- 10
lambda<- 2

#specifying seed
set.seed(32114)

#generating N(t)
njumps<- rpois(1,lambda*t)
```

```
#defining states
N<- 0:njumps

#generating N(t) standard uniforms
u<- c()
u[1]<- 0

for(i in 2:(njumps+1))
u[i]<- runif(1)

#computing event times
time<- t*sort(u)

#plotting trajectory
plot(time, N, type="n", xlab="Time", ylab="State",
panel.first = grid())

segments(time[-length(time)], N[-length(time)],
time[-1]-0.07, N[-length(time)], lwd=2, col="blue")

points(time, N, pch=20, col="blue")
points(time[-1], N[-length(time)], pch=1, col="blue")
```

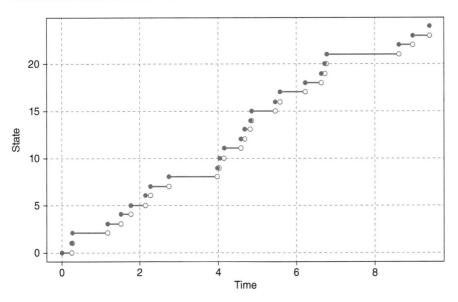

□

3.3   Applications of Poisson Process

APPLICATION 3.1. In seismology, occurrence of earthquakes is often modeled according to a Poisson process. We obtain the data from the Southern California Earthquake Data Center's website *https://service.scedc.caltech.edu/ eq-catalogs/date_mag_loc.php*. The data are on earthquakes in Southern California with a minimum magnitude of 3.0 that occurred between 2012 and 2018. We compute the lengths of the interarrival times and remove those earthquakes that were registered within three hours of their predecessors (possibly aftershocks). Finally, we conducted a chi-squared goodness-of-fit test to see if these times follow an exponential distribution. The R code and output follow.

```
eq.data<- read.csv(file="./earthquakedata2012-2018.csv",
header=TRUE, sep=",")

#creating date-time variable
datetime<- as.POSIXct(paste(as.Date(eq.data$DATE),
eq.data$TIME))

#computing lag
datetime.lag<- c(0,head(datetime, -1))

#computing interarrival times (in hours)
int.time<- (as.numeric(datetime)-as.numeric(datetime.lag))/
3600

#removing first value
int.time<- int.time[-1]

#removing immediate aftershocks (within 3 hours)
int<- int.time[int.time>3]

#plotting histogram
hist(int, main="", col="dark magenta", xlab="Interarrival
Time")
```

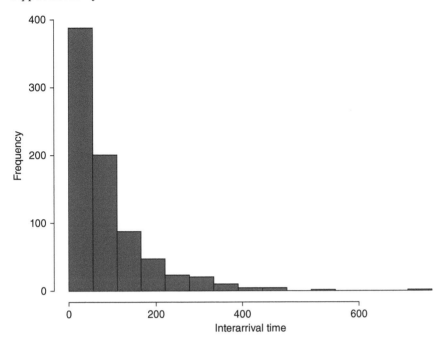

```
#binning interarrival times
binned.int<- as.factor(ifelse(int<40,"1",
ifelse(int>=40 & int<80,"2",ifelse(int>=80 & int<120,3",
ifelse(int>=120 & int<160,"4",ifelse(int>=160 &
int<200,"5",
ifelse(int>=200 & int<240,"6","7")))))))

#computing observed frequencies
obs<- table(binned.int)

#estimating mean for exponential distribution
mean.est<- mean(int)

#computing expected frequencies
exp<- c(1:7)
exp[1]<- length(int)*(1-exp(-40/mean.est))
exp[2]<- length(int)*(exp(-40/mean.est)-exp(-80/mean.est))
exp[3]<- length(int)*(exp(-80/mean.est)-exp(-120/mean.est))
exp[4]<- length(int)*(exp(-120/mean.est)-exp(-160/mean.est))
exp[5]<- length(int)*(exp(-160/mean.est)-exp(-200/mean.est))
exp[6]<- length(int)*(exp(-200/mean.est)-exp(-240/mean.est))
exp[7]<- length(int)*exp(-240/mean.est)

obs
```

```
  1    2    3    4    5    6    7
342  178  117   49   39   24   42
```

```
round(exp,1)
```

```
319.8 190.5 113.5  67.6  40.3  24.0  35.4
```

```
#computing chi-squared statistic
print(chi.sq<- sum((obs-exp)^2/exp))
```

```
8.883823
```

```
#computing p-value
print(p.value<- 1-pchisq(chi.sq, df=5))
```

```
0.1137888
```

The number of degrees of freedom in this test is calculated as the number of bins minus 1 and minus the number of parameters that have to be estimated (in this case one mean), so we get that $df = 7-1-1 = 5$. The p-value is larger than 0.05, indicating that the earthquakes in the given time frame occurred according to a Poisson process. □

APPLICATION 3.2. In sports analytics, a Poisson process is used to model the process of goal scoring in a game. Consider a team game where players score only one point at a time, for instance, ice hockey. Suppose the points scored by team A follow a Poisson process $\{N_A(t),\, t \geq 0\}$ with rate λ_A, and points scored by team B are governed by a Poisson process $\{N_B(t),\, t \geq 0\}$ with parameter λ_B. Assuming the two processes are independent, we can derive some interesting results.

(a) The sum of the two independent Poisson processes $N(t) = N_A(t) + N_B(t)$ is the superposition Poisson process with rate $\lambda_A + \lambda_B$. It represents the process of scoring by either team A or B. If on average, fans wait for time $1/\lambda_A$ for team A to score, respectively, $1/\lambda_B$ for team B to score, then the officials wait, on average, for a shorter period of time $1/(\lambda_A + \lambda_B)$ for either team to score.

To see how it works with numbers, suppose team A scores, on average, every 10 minutes, and team B scores every 12 minutes, on average. Then the expected waiting time until any team scores is $1/(1/10+1/12) = 120/22 = 5.45$ seconds.

(b) We can find the probability that one team scores ahead of the other team.

Denote by T_A and T_B the respective interarrival times. We know that T_A and T_B are independent and exponentially distributed with means $\mathbb{E}(T_B) = 1/\lambda_A$ and $\mathbb{E}(T_B) = 1/\lambda_B$. We write

$$\mathbb{P}(\text{team } A \text{ scores before team } B) = \mathbb{P}(T_A < T_B)$$

$$= \int_0^\infty \mathbb{P}(T_B > t)\, f_A(t)\, dt = \int_0^\infty e^{-\lambda_B t}\, \lambda_A\, e^{-\lambda_A t}\, dt = \frac{\lambda_A}{\lambda_A + \lambda_B}.$$

Now switching λ_A and λ_B, we get

$$\mathbb{P}(\text{team } B \text{ scores before team } A) = \mathbb{P}(T_B < T_A) = \frac{\lambda_B}{\lambda_A + \lambda_B}.$$

With our numbers, $\mathbb{P}(\text{team } A \text{ scores before } B) = \frac{1/10}{1/10+1/12} = 12/22 = 0.545$, and $\mathbb{P}(\text{team } B \text{ scores before } A) = 1 - 0.545 = 0.455$.

(c) We can find the probability of a tie at the end of the game, and also the probability that team A (team B) wins.

Let T denote the length of the game. Using independence of the two processes, $\{N_A(t),\ t \geq 0\}$ and $\{N_B(t),\ t \geq 0\}$, and expressions for the probability mass functions, we write

$$\mathbb{P}(\text{game ties}) = \sum_{n=0}^{\infty} \mathbb{P}(N_A(T) = n,\ N_B(T) = n)$$

$$= \sum_{n=0}^{\infty} \mathbb{P}(N_A(T) = n)\, \mathbb{P}(N_B(T) = n)$$

$$= \sum_{n=0}^{\infty} \frac{(\lambda_A T)^n}{n!} e^{-\lambda_A T} \cdot \frac{(\lambda_B T)^n}{n!} e^{-\lambda_B T} = e^{-(\lambda_A+\lambda_B)T} \sum_{n=0}^{\infty} \frac{(\lambda_A \lambda_B T^2)^n}{(n!)^2},$$

$$\mathbb{P}(\text{team } A \text{ wins}) = \mathbb{P}(N_A(T) > N_B(T))$$

$$= \sum_{n=0}^{\infty} \sum_{k=1}^{\infty} \mathbb{P}(N_A(T) = n+k,\ N_B(T) = n)$$

$$= \sum_{n=0}^{\infty} \sum_{k=1}^{\infty} \mathbb{P}(N_A(T) = n+k)\, \mathbb{P}(N_B(T) = n)$$

$$= \sum_{n=0}^{\infty} \sum_{k=1}^{\infty} \frac{(\lambda_A T)^{n+k}}{(n+k)!} e^{-\lambda_A T} \cdot \frac{(\lambda_B T)^n}{n!} e^{-\lambda_B T}$$

$$= e^{-(\lambda_A+\lambda_B)T} \sum_{n=0}^{\infty} \left[\frac{(\lambda_A\lambda_B T^2)^n}{n!} \cdot \sum_{k=1}^{\infty} \frac{(\lambda_A T)^k}{(n+k)!} \right],$$

and, switching λ_A and λ_B, we get

$$\mathbb{P}(\text{team } B \text{ wins}) = e^{-(\lambda_A + \lambda_B)T} \sum_{n=0}^{\infty} \left[\frac{(\lambda_A \lambda_B T^2)^n}{n!} \cdot \sum_{k=1}^{\infty} \frac{(\lambda_B T)^k}{(n+k)!} \right].$$

The duration of the playing time in an ice hockey game is $T = 60$ minutes. Therefore, we obtain

$$\mathbb{P}(\text{game ties}) = e^{-(1/10+1/12)(60)} \sum_{n=0}^{\infty} \frac{\left((1/10)(1/12)(60)^2\right)^n}{(n!)^2}$$

$$= e^{-11} \sum_{n=0}^{\infty} \frac{30^n}{(n!)^2} = 0.1166.$$

We calculated the sum numerically in R. The sum converges after 15 terms.

```
sum<- 0
for(n in 0:15)
sum<- sum+30^n/(factorial(n))^2

sum*exp(-11)
```

0.1165575

Further,

$$\mathbb{P}(\text{team } A \text{ wins}) = e^{-11} \sum_{n=0}^{\infty} \left[\frac{30^n}{n!} \cdot \sum_{k=1}^{\infty} \frac{6^k}{(n+k)!} \right] = 0.5590.$$

R code given below computes this double sum numerically.

```
sum.n<- 0
for (n in 0:15) {
sum.k<-0
   for (k in 1:15)
     sum.k<- sum.k+6^k/factorial(n+k)
sum.n<- sum.n + 30^n/factorial(n)*sum.k
}

sum.n*exp(-11)
```

0.5589743

Finally,

$$\mathbb{P}(\text{team } B \text{ wins}) = e^{-11} \sum_{n=0}^{\infty} \left[\frac{30^n}{n!} \cdot \sum_{k=1}^{\infty} \frac{5^k}{(n+k)!} \right] = 0.3244,$$

as the R code below computes

```
sum.n<- 0
for (n in 0:15) {
sum.k<-0
   for (k in 1:15)
     sum.k<- sum.k+5^k/factorial(n+k)
sum.n<- sum.n + 30^n/factorial(n)*sum.k
}

sum.n*exp(-11)
```

0.3244495

Note that the three probabilities add up to 1, as they should be. □

APPLICATION 3.3. One famous application of a Poisson process is that of a pedestrian versus traffic flow. A pedestrian needs to get to the other side of a road. Assume that cars pass according to a Poisson process with rate λ.

(a) If it takes the pedestrian time τ to cross the road, how long, on average, will it take the person to get to the other side?

Note that the person has to wait for a gap in traffic of length at least τ before he/she can cross. Denote by T the total time (waiting plus crossing). Suppose the person just approached the road. The Poisson process renews itself at this moment, and thus, the person has to wait an exponential time with a mean $1/\lambda$ for the next car. Call this time T_1. If $T_1 \geq \tau$, then the person can safely cross the road and $T = \tau$. If, however, $T_1 < \tau$, the person has to wait for T_1 for the first car to pass, and then the process renews itself and the pedestrian would have to wait for an additional time $\tilde{T}$ that has the same distribution at T. Thus, T can be written as

$$T = \begin{cases} \tau, & \text{if } T_1 \geq \tau, \\ T_1 + \tilde{T}, & \text{if } T_1 < \tau, \end{cases}$$

where T_1 is an exponentially distributed random variable with mean $1/\lambda$. Consequently,

$$\mathbb{E}(T) = \tau \mathbb{P}(T_1 \geq \tau) + \int_0^\tau t\lambda e^{-\lambda t}\, dt + \mathbb{E}(T)\mathbb{P}(T_1 < \tau).$$

This can be rewritten as

$$\mathbb{E}(T)\,\mathbb{P}(T_1 \geq \tau) = \tau\,\mathbb{P}(T_1 \geq \tau) + \int_0^\tau t\lambda e^{-\lambda t}\,dt$$

$$= \tau e^{-\lambda\tau} - \tau e^{-\lambda\tau} + \int_0^\tau e^{-\lambda t}\,dt = \frac{1}{\lambda}\left(1 - e^{-\lambda\tau}\right).$$

From here,

$$\mathbb{E}(T) = \frac{1 - e^{-\lambda\tau}}{\lambda e^{-\lambda\tau}} = \frac{1}{\lambda}(e^{\lambda\tau} - 1).$$

Suppose, on average, a car passes every 20 seconds (that is, $\lambda = 1/20 = 0.05$ cars per second), and the pedestrian needs $\tau = 30$ seconds to cross the road. Thus, on average, it takes the person $\mathbb{E}(T) = \frac{1}{0.05}\left(e^{(0.05)(30)} - 1\right) = 69.63$ seconds to cross the road.

(b) How many cars, on average, will pass by before the pedestrian can cross?

Let N be the number of cars that pass by before the person can cross. Then $N \geq n$, if and only if the first n interarrival times are all less than τ. Hence, $\mathbb{P}(N \geq n) = \left(1 - e^{-\lambda\tau}\right)^n$. We can compute the probability of $N = n$ as

$$\mathbb{P}(N = n) = \mathbb{P}(N \geq n) - \mathbb{P}(N \geq n+1) = \left(1 - e^{-\lambda\tau}\right)^n - \left(1 - e^{-\lambda\tau}\right)^{n+1}.$$

Let $a = 1 - e^{-\lambda\tau}$. The expected value of N can then be computed as

$$\mathbb{E}(N) = \sum_{n=0}^\infty n\,\mathbb{P}(N = n) = \sum_{n=0}^\infty n\left(a^n - a^{n+1}\right) = a - a^2 + (2)(a^2 - a^3)$$

$$+(3)(a^3 - a^4) + \cdots = a + a^2 + a^3 + \cdots = \frac{1}{1-a} - 1 = e^{\lambda\tau} - 1.$$

Note that $\mathbb{E}(T) = (1/\lambda)\mathbb{E}(N)$. It is intuitively so, because the person has to wait until an average of $\mathbb{E}(N)$ cars pass, and the average waiting time between these cars is $1/\lambda$ seconds. In our numeric example, $\lambda = 0.05$ and $\tau = 30$. So, $\mathbb{E}(N) = e^{(0.05)(30)} - 1 = 3.48$ cars.   □

Exercises

EXERCISE 3.1. Let $\{N(t),\, t \geq 0\}$ be a Poisson process with rate λ. Find the joint probability distribution $\mathbb{P}(N(s) = m,\, N(t) = n)$, for any $t \geq s \geq 0$, and $n \geq m \geq 0$.

EXERCISE 3.2. Show that for a Poisson process $\{N(t), t \geq 0\}$ with rate λ, the covariance between $N(s)$ and $N(t)$ is equal to $\lambda \min(s, t)$, for any $s, t \geq 0$.

EXERCISE 3.3. An insurance agent handles policyholders' claims. Claims are submitted on weekdays according to a Poisson process with a rate $\lambda = 5$ per day.
(a) If there were two claims submitted on Monday and three on Tuesday, what is the probability that by the end of the day on Friday there will be a total of 16 claims submitted that week?
(b) In the new calendar year, the agent opens for business on Monday, January 2. On what day does he expect to see the 100th claim?

EXERCISE 3.4. A salesperson contacts customers over the phone and offers his product. Assume that the times that pass between consecutive phone calls (that includes the call and the break in-between) are independent and exponentially distributed with mean of 5 minutes. He estimates that 15% of all the customers he calls actually buy his product.
(a) Calculate the expected number of successful sales in the next 2 hours.
(b) Compute the probability that within 1 hour he places 15 calls, 5 of which result in a sale.
(c) Find the conditional probability that he makes 10 sales in 4 hours, given that he has made 3 sales the first hour.

EXERCISE 3.5. People contract a disease according to a Poisson process with an unknown rate λ. Suppose the incubation period until symptoms of the disease show is a random variable with a known cumulative distribution function F. Let $N_1(t)$ denote the number of individuals who have shown symptoms by time t, and let $N_2(t)$ be the number of individuals who have not yet shown any symptoms by time t.
(a) Argue that $\{N_1(t), t \geq 0\}$ and $\{N_2(t), t \geq 0\}$ are independent Poisson processes with means

$$\mathbb{E}(N_1(t)) = \lambda \int_0^t F(u)\, du \ \text{ and } \ \mathbb{E}(N_2(t)) = \lambda \int_0^t (1 - F(u))\, du.$$

(b) For a known time t and observed number of individuals showing symptoms $\widehat{\mathbb{E}}(N_1(t))$, prove that the estimated number of individuals infected but not yet showing symptoms is

$$\widehat{\mathbb{E}}(N_2(t)) = \frac{\widehat{\mathbb{E}}(N_1(t)) \int_0^t (1 - F(u))\, du}{\int_0^t F(u)\, du}.$$

(c) Suppose the incubation period until symptoms show is an exponentially distributed random variable with a mean of 2 days. If 1,000 individuals show symptoms of a disease by day 10, estimate the number of individuals who are also infected but haven't shown the symptoms yet.

EXERCISE 3.6. Areas of high road surface distress (potholes or cracks) that need immediate attention of road maintenance operators are distributed according to a Poisson law with a rate of 2.8 per mile.
(a) What is the average number of distressed road surface areas on a 10-mile stretch of a freeway?
(b) Simulate locations of 30 distressed surface areas. What is the total length of the road in your simulation?
(c) Suppose there are 30 distressed surface areas on a 10-mile stretch of a freeway. Simulate locations of those areas.

EXERCISE 3.7. The National Geophysical Data Center's website

https://www.ngdc.noaa.gov/hazel/view/hazards/volcano/event-search/

provides access to the Global Significant Volcanic Eruptions Database. Verify that those volcanic eruptions in the past 100 years can be modeled as a Poisson process.

EXERCISE 3.8. Two teams are playing basketball. Team A opportunities to score appear as a Poisson process with a rate of 0.5 per minute. Suppose that 25% bring one point, 40% bring two points, 20% bring the team three points, and the others result in missed shots. For team B, the opportunities come as a Poisson process with a rate of 0.4 per minute, of which 25% result in a 1-pointer, 50% result in a 2-pointer, 15% result in a 3-pointer, and the rest are missed.
(a) How long, on average, does the stadium have to wait until a team scores?
(b) How long, on average, will the fans wait until team A scores? Team B scores?
(c) What is the probability that team A scores before team B? Team B scores before team A?
(d) What is the probability that at the end of the 48-minute game, teams A and B will score the same number of 1-pointers, the same number of 2-pointers, and the same number of 3-pointers?

EXERCISE 3.9. In a popular nursery rhyme,

Itsy bitsy spider went up the water spout.
Down came the rain and washed the spider out.
Out came the sun and dried up all the rain,
And the itsy bitsy spider went up the spout again.

Assume that the length of the downspout is 30 feet, and the spider climbs with a constant speed of 1 foot per minute. The rain comes down as a Poisson process with a rate of 2 per hour.

(a) Find the expected time it takes the spider to reach the top.

(b) Find the expected number of times the spider will be washed down before it reaches the top.

4

Nonhomogeneous Poisson Process

4.1 Definition of Nonhomogeneous Poisson Process

The Poisson process considered in the previous chapter has a constant rate λ. In some situations, it is difficult to assume that the rate doesn't change over a large period of time. In this case, we can make λ depend on time t and define a Poisson process with rate $\lambda(t)$. The rate is now called the *intensity rate* or *intensity function*. The process still starts at zero at time zero, and its increments are still independent, but now the increments are non-stationary.

A counting process $\{N(t), t \geq 0\}$ is called a *nonhomogeneous* (or *non-stationary*, or *time-dependent*) Poisson process[1] if: (i) $N(0) = 0$, (ii) increments are independent, and (iii) for all $s, t \geq 0$,

$$\mathbb{P}\big(N(t+s) - N(s) = n\big) = \frac{\left(\int_s^{t+s} \lambda(u)\, du \right)^n}{n!} \, e^{-\left(\int_s^{t+s} \lambda(u)\, du \right)}, \quad n \geq 0.$$

Define the function $\Lambda(t) = \int_0^t \lambda(u)\, du$. It is called the *integrated intensity rate function* or the *mean value function*. The probability mass function of a nonhomogeneous Poisson process can be written in terms of $\Lambda(t)$ as

$$\mathbb{P}\big(N(t+s) - N(s) = n\big) = \frac{[\Lambda(t+s) - \Lambda(s)]^n}{n!} \, e^{-[\Lambda(t+s) - \Lambda(s)]}, \quad n \geq 0.$$

Note that for $s, t \geq 0$, $\mathbb{E}(N(t+s) - N(s)) = \Lambda(t+s) - \Lambda(s)$.

REMARK 4.1. Here we formulate an alternative definition of a nonhomogeneous Poisson process. It can be shown that the two definitions are equivalent. We will need this definition to justify the method we use in Simulation 4.3. A counting process $\{N(t), t \geq 0\}$ is termed a *nonhomogeneous* Poisson process with the intensity rate $\lambda(t)$, $t \geq 0$, if: (i) $N(0) = 0$, (ii) increments are

---

[1] Introduced by a prominent statistician Sir David Roxbee Cox in his 1955 paper "Some Statistical Methods Connected with Series of Events." *Journal of the Royal Statistical Society. Series B (Methodological)*, 17(2): 129 – 164.

DOI: 10.1201/9781003244288-4

independent and stationary, (iii) for any fixed $t \geq 0$ and any small positive increment Δt, $\mathbb{P}(\text{no events happen in } [t, t+\Delta t]) = 1 - \lambda(\Delta t) + o(\Delta t)$, and (iv) $\mathbb{P}(\text{one event happens in } [t, t+\Delta t]) = \lambda(\Delta t) + o(\Delta t)$, where $o(\Delta t)$ (pronounced "little oh of delta t") denotes any function of Δt that goes to zero faster than Δt. That is, $o(\Delta t) = \{f(\Delta t) : \lim_{\Delta t \to 0} \frac{f(\Delta t)}{\Delta t} = 0\}$. $\square$

EXAMPLE 4.1. Throughout the day, arrivals of phone calls to a doctor's office can be modeled as a nonhomogeneous Poisson process with the intensity rate

$$\lambda(t) = \begin{cases} 10, & \text{if 9AM} \leq t \leq \text{10:30AM}, \\ 5, & \text{if 10:30AM} < t \leq \text{12PM}, \\ 8, & \text{if 12PM} < t \leq \text{1PM}, \\ 4, & \text{if 1PM} < t \leq \text{5PM}. \end{cases}$$

(a) To calculate the integrated rate function $\Lambda(t)$, we need to define the time variable as ranging between 0 and 8 hours of the workday. We can rewrite the intensity rate as

$$\lambda(t) = \begin{cases} 10, & \text{if } 0 \leq t \leq 1.5, \\ 5, & \text{if } 1.5 < t \leq 3, \\ 8, & \text{if } 3 < t \leq 4, \\ 4, & \text{if } 4 < t \leq 8. \end{cases}$$

The integrated rate function is then computed as

$$\Lambda(t) = \int_0^t \lambda(u)\, du = \begin{cases} \int_0^t 10\, du = 10t, & \text{if } 0 \leq t \leq 1.5, \\ 15 + \int_{1.5}^t 5\, du = 15 + 5(t-1.5), & \text{if } 1.5 < t \leq 3, \\ 22.5 + \int_3^t 8\, du = 22.5 + 8(t-3), & \text{if } 3 < t \leq 4, \\ 30.5 + \int_4^t 4\, du = 30.5 + 4(t-4), & \text{if } 4 < t \leq 8. \end{cases}$$

(b) Below we plot both functions. The R codes are provided.

```
#plotting intensity rate
t=c(0,1.5,3,4,8)
lambda=c(10, 10,5,8,4)
plot(t, lambda, type="n", col="blue ", xlim=c(0,8),
ylim=c(0,12), xlab="Time", ylab="Intensity rate")
segments(t[-5]+0.07, lambda[-1], t[-1], lambda[-1], lwd=2,
col="blue")
points(t, lambda, cex=1.2, pch=19, col="blue")
points(t[-5], lambda[-1], cex=1.2, pch=1, col="blue")
```

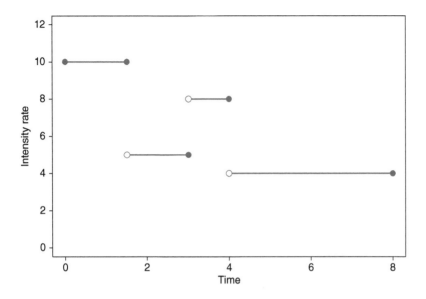

```
#plotting integrated rate function
t<- c(0, 1.5, 3, 4, 8)
Lambda<- c(0, 15, 22.5, 30.5, 46.5)
plot(t,Lambda, type="l", lwd=2, col="blue", xlim=c(0,8),
ylim=c(0,50), xlab="time", ylab="integrated rate function")
points(t, Lambda, cex=1.2, pch=16, col="blue")
```

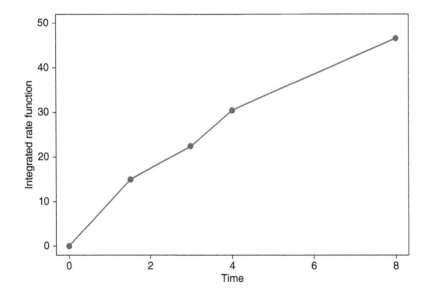

(c) Suppose we want to compute the probability that there will be 15 phone calls between 11AM and 2PM. The time 11AM corresponds to 2 hours and 2PM corresponds to 5 hours after the office opens. We write

$$\mathbb{P}\big(N(5) - N(2) = 15\big) = \frac{\big[\Lambda(5) - \Lambda(2)\big]^{15}}{15!} \, e^{-\big[\Lambda(5) - \Lambda(2)\big]}$$

$$= \frac{\big(30.5 + 4(5-4) - (15 + 5(2-1.5))\big)^{15}}{15!} \, e^{-\big(30.5 + 4(5-4) - (15 + 5(2-1.5))\big)}$$

$$= \frac{(17)^{15}}{15!} \, e^{-17} = 0.011468.$$

(d) Finally, we want to compute the average number of phone calls per day. We compute

$$\mathbb{E}\big(N(8) - N(0)\big) = \Lambda(8) - \Lambda(0) = 30.5 + 4(8-4) - 0 = 46.5 \ \text{ calls.} \quad \square$$

4.2   Simulations in R

Recall that in Section 3.2 we discussed two simulation methods for trajectories of a homogeneous Poisson process. In the present section, we generalize these methods to the case of a nonhomogeneous Poisson process.

SIMULATION 4.1. (EXPONENTIAL INTERARRIVALS). In this method, we simulate interarrival times. In the nonhomogeneous case, the distributions of interarrival times are not independent. They are obtained as follows.

The first interarrival time has the cumulative distribution function $F_{T_1}(t) = 1 - e^{-\Lambda(t)}$, $t \geq 0$. For a fixed time of the first event occurrence $S_1 = T_1 = s_1$, the cumulative distribution function of T_2 is

$$F_{T_2 \mid S_1}(t \mid s_1) = 1 - e^{-\big(\Lambda(t + s_1) - \Lambda(s_1)\big)}, \ t \geq 0.$$

In general, for a given waiting time for the nth event $S_n = s_n$, the conditional distribution of the interarrival time T_{n+1} is

$$F_{T_{n+1} \mid S_n}(t \mid s_n) = 1 - e^{-\big(\Lambda(t + s_n) - \Lambda(s_n)\big)}, \ t \geq 0, \ n \geq 1.$$

As an example, here we give the code that simulates a trajectory of a nonhomogeneous Poisson process with the integrated rate function $\Lambda(t) = t + 0.05\,t^2$, $t \geq 0$. In the code we first generate standard uniform random variables $U_i, i = 1, \ldots, n$, and then compute event times by inverting the cumulative distribution function. We write $1 - e^{-(S_1 + 0.05\,S_1^2)} = U_1$, which solution is $S_1 = \sqrt{100 - 20 \ln(1 - U_1)} - 10$. It can be simplified by replacing $1 - U_1$ by U_1 since both are standard uniform random variables. Thus we have $S_1 = \sqrt{100 - 20 \ln(U_1)} - 10$.

The second event time S_2 solves $1 - e^{-\left[S_2 + 0.05\, S_2^2 - (S_1 + 0.05\, S_1^2) \right]} = U_2$, or equivalently, $(S_2 + 10)^2 - (S_1 + 10)^2 = -20 \ln(1 - U_2)$. The solution is (with $1 - U_2$ replaced by U_2) $S_2 = \sqrt{(S_1 + 10)^2 - 20 \ln(U_2)} - 10$. The general recurrence formula is $S_{n+1} = \sqrt{(S_n + 10)^2 - 20 \ln(U_{n+1})} - 10$. We continue generating the event times according to this formula until we reach a pre-specified number of events of the process. The code and graph are given below.

```
#specifying parameters
njumps<- 20

#defining states
N<- 0:njumps

#defining times as vectors
time<- c()

#specifying seed
set.seed(76855)

#generating standard uniforms
u<- c()
for(i in 1:njumps)
u[i]<- runif(1)

#computing event times
time[1]<- 0
time[2]<- sqrt(100-20*log(u[1]))-10

for(i in 3:(njumps+1)) {
time[i]<- sqrt((time[i-1]+10)^2-20*log(u[i-1]))-10
}

#plotting trajectory
plot(time, N, type="n", xlab="Time", ylab="State",
panel.first=grid())

segments(time[-length(time)], N[-length(time)],
time[-1]-0.07, N[-length(time)], lwd=2, col="blue")

points(time, N, pch=20, col="blue")
points(time[-1], N[-length(time)], pch=1, col="blue")
```

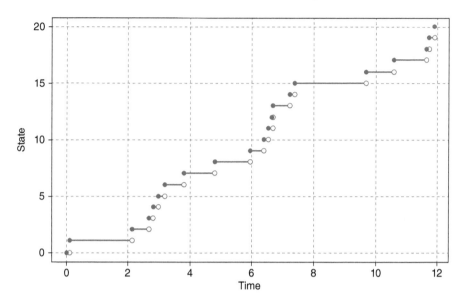

□

SIMULATION 4.2. (UNIFORM ORDER STATISTICS). Let $\{N(t),\ t \geq 0\}$ denote a nonhomogeneous Poisson process with the intensity rate $\lambda(t)$, $t \geq 0$, and integrated rate function $\Lambda(t)$, $t \geq 0$. Given that n events occurred by time t, the conditional joint density of n waiting times $S_1, S_2, \ldots, S_n$, is derived as

$$f_{S_1, S_2, \ldots, S_n}\left(s_1, s_2, \ldots, s_n \mid N(t) = n\right)$$

$$= \frac{f_{T_1}(s_1) f_{T_2}(s_2 - s_1) \cdots \cdot f_{T_n}(s_n - s_{n-1}) \mathbb{P}(T_{n+1} > t - s_n)}{\mathbb{P}(N(t) = n)}$$

$$= \frac{\lambda(s_1)\, e^{-\Lambda(s_1)}\, \lambda(s_2 - s_1)\, e^{-(\Lambda(s_2) - \Lambda(s_1))} \cdots \lambda(s_n - s_{n-1})}{\dfrac{(\Lambda(t))^n}{n!}\, e^{-\Lambda(t)}}$$

$$= n!\, \frac{\lambda(s_1)\, \lambda(s_2 - s_1) \cdots \cdot \lambda(s_n - s_{n-1})}{\Lambda(t)^n}.$$

This means that the event times are order statistics from the distribution with density $f_S(s) = \dfrac{\lambda(s)}{\Lambda(t)}$, $0 \leq s \leq t$, and the cumulative distribution function $F_S(s) = \dfrac{\Lambda(s)}{\Lambda(t)}$, $0 \leq s \leq t$.

In our example with $\Lambda(t) = t + 0.05\, t^2$, $t \geq 0$, to generate a trajectory, we proceed as follows.

Step 1. Fix t and generate $N(t) \sim Poi(\Lambda(t))$. For instance, for $t = 10$, $\Lambda(10) = 10 + (0.05)(10^2) = 15$, and so, we would generate $N(t) \sim Poi(15)$.
Step 2. Generate $N(t)$ standard uniform random variables $U_1, \ldots, U_{N(t)}$.
Step 3. Order $U_1, \ldots, U_{N(t)}$ in increasing order, obtaining the ordered set $U_{(1)}, \ldots, U_{(N(t))}$.
Step 4. Compute waiting times $S_1, \ldots, S_{N(t)}$ that are positive solutions of the equations $\dfrac{\Lambda(S_i)}{\Lambda(t)} = U_{(i)}$, or $S_i + 0.05\,S_i^2 = 15\,U_{(i)}$. That is, $S_i = 10\sqrt{1 + 3\,U_{(i)}} - 10$.
Step 5. Define the states of the Poisson process as $N(0) = 0, N(S_1) = 1, N(S_2) = 2, \ldots, N(S_{N(t)}) = N(t)$.
Step 6. Plot the states against time.

The code and plot follow.

```
#specifying parameters
t<- 10
Lambda<- t+0.05*t^2

#specifying seed
set.seed(997755)

#generating N(t)
njumps<- rpois(1,Lambda)

#defining states
N<- 0:njumps

#generating N(t) standard uniforms
u<- c()
u[1]<- 0

for(i in 2:(njumps+1))
u[i]<- runif(1)

#computing event times
time<- 10*sqrt(1+3*sort(u))-10

#plotting trajectory
plot(time, N, type="n", xlab="Time", ylab="State",
panel.first=grid())
```

```
segments(time[-length(time)], N[-length(time)],
time[-1]-0.07, N[-length(time)], lwd=2, col="blue")

points(time, N, pch=20, col="blue")
points(time[-1], N[-length(time)], pch=1, col="blue")
```

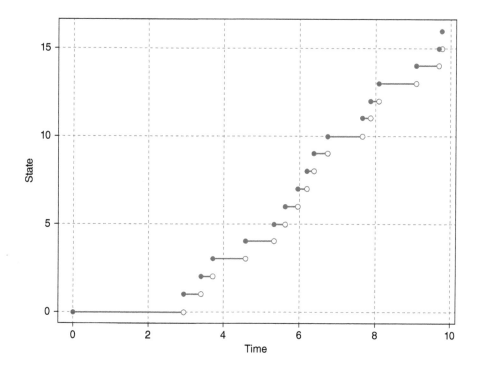

□

Below we introduce a third method of simulating a trajectory of a nonhomogeneous Poisson process. In the case of a homogeneous Poisson process, this method becomes trivial and hence not useful.

SIMULATION 4.3. (THINNING). Let $\{N(t),\ t \geq 0\}$ denote a nonhomogeneous Poisson process with the intensity rate $\lambda(t)$, $t \geq 0$, and suppose its integrated rate function doesn't have an explicit form or is not easily invertible, so the previous two simulation methods won't work for this process. Here we introduce another method, called the *thinning* method. In this method, first we generate a nonhomogeneous process $\{N^*(t),\ t \geq 0\}$ with intensity rate $\lambda^*(t)$, $t \geq 0$, that uniformly dominates $\lambda(t)$. That is, $\lambda(t) \leq \lambda^*(t)$ for all $t \geq 0$. The process is selected in such a way that its integrated rate function is invertible, and so $N^*(t)$ can be generated by either of the previous two methods.

Sometimes $\lambda^*(t)$ may be chosen to be a constant, resulting in a homogeneous Poisson process, which is even simpler to generate (Section 3.2).

Further, once the event times $s_1^*, ..., s_N^*$ for the process $\{N^*(t), t \geq 0\}$ are generated, they are "thinned" according to the following acceptance-rejection rule: if a standard uniform random variable is less than or equal to the ratio $\lambda(s_i^*)/\lambda^*(s_i^*)$, the time s_i^* is accepted, otherwise, rejected.

The accepted times are the event times for the process $\{N(t), t \geq 0\}$. To show this, we use the alternative definition of a nonhomogeneous Poisson process given in Remark 4.1 and argue as follows. The process $\{N^*(t), t \geq 0\}$ starts at zero and has independent and stationary increments. These properties are inherited by $\{N(t), t \geq 0\}$. Further, in an infinitesimally small interval of length Δt, there are zero occurrences of the process $\{N(t), t \geq 0\}$ if there are no occurrences of the process $\{N^*(t), t \geq 0\}$ or there is one occurrence but it is rejected. There is one occurrence of the process $\{N(t), t \geq 0\}$ if there is one occurrence of the process $\{N^*(t), t \geq 0\}$ and it is accepted. Denoting by T^* an interarrival time, and by U a standard uniform random variable, we write

$$\mathbb{P}\big(N(\Delta t) = 0\big) = \mathbb{P}\big(N^*(\Delta t) = 0\big) + \mathbb{P}\big(N^*(\Delta t) = 1\big)\,\mathbb{P}\Big(U > \frac{\lambda(\Delta t)}{\lambda^*(\Delta t)}\Big)$$

$$= \mathbb{P}\big(T^* > \Delta t\big) + \mathbb{P}\big(T^* < \Delta t\big)\,\mathbb{P}\Big(U > \frac{\lambda(\Delta t)}{\lambda^*(\Delta t)}\Big)$$

$$= e^{-\lambda^*(\Delta t)} + \big(1 - e^{-\lambda^*(\Delta t)}\big)\Big(1 - \frac{\lambda(\Delta t)}{\lambda^*(\Delta t)}\Big)$$

$$= 1 - \lambda^*(\Delta t) + o(\Delta t) + \big(\lambda^*(\Delta t) + o(\Delta t)\big)\Big(1 - \frac{\lambda(\Delta t)}{\lambda^*(\Delta t)}\Big)$$

$$= 1 - \lambda(\Delta t) + o(\Delta t), \quad \text{for small } \Delta t,$$

and

$$\mathbb{P}\big(N(\Delta t) = 1\big) = \mathbb{P}\big(N^*(\Delta t) = 1\big)\,\mathbb{P}\Big(U \leq \frac{\lambda(\Delta t)}{\lambda^*(\Delta t)}\Big)$$

$$= \mathbb{P}\big(T^* < \Delta t\big)\,\mathbb{P}\Big(U \leq \frac{\lambda(\Delta t)}{\lambda^*(\Delta t)}\Big) = \big(1 - e^{-\lambda^*(\Delta t)}\big)\frac{\lambda(\Delta t)}{\lambda^*(\Delta t)}$$

$$= \big(\lambda^*(\Delta t) + o(\Delta t)\big)\frac{\lambda(\Delta t)}{\lambda^*(\Delta t)} = \lambda(\Delta t) + o(\Delta t), \quad \text{for small } \Delta t.$$

The above derivation shows that $N(t)$ is a nonhomogeneous Poisson process with the intensity rate function $\lambda(t)$.

Below we present the code that simulates a trajectory of a nonhomogeneous Poisson process with the intensity rate $\lambda(t) = 20 + 20\sin(\pi t)$, $t \geq 0$. For this process, the integrated intensity rate function $\Lambda(t) = 20t - \frac{20}{\pi}\cos(\pi t) + \frac{20}{\pi}$, $t \geq 0$, is not readily invertible. However, $\lambda(t) \leq 40$, and so we simulate event times for a Poisson process with rate $\lambda^*(t) = 40$, and then accept an event time s if $U \leq \lambda(s)/\lambda^*(s) = 20(1 + \sin(\pi s))/40 = 0.5(1 + \sin(\pi s))$, and reject otherwise.

```
#specifying parameters
lambda<- function(t) { 20+20*sin(pi*t) }
lambda.star<- function(t) 40
Lambda.star<- function(t) 40*t

#specifying seed
set.seed(2866514)

#generating N(10)
njumps<- rpois(1, Lambda.star(10))

#generating N(10) standard uniforms
u<- c()
u[1]<- 0

for(i in 2:(njumps+1))
u[i]<- runif(1)

#computing event times
time.star<- 10*sort(u)

#thinning event times
accepted<- c()
time<- c()
accepted[1]<- 1
time[1]<- 0

for (i in 2:(njumps+1)) {
if (runif(1)<= lambda(time.star[i])/lambda.star(time.star[i]))
accepted[i]=1 else accepted[i]=0
}

time<- time.star[-which(accepted==0)]
N<- 0:(length(time)-1)

#plotting trajectory
plot(time, N, type="n", xlab="Time", ylab="State",
panel.first = grid())
```

```
segments(time[-length(time)],N[-length(time)], time[-1]-0.07,
N[-length(time)], lwd=2, col="blue")

points(time, N, ylim=c(0,120), pch=20, col="blue")
points(time[-1],N[-length(time)],pch=1, col="blue")
```

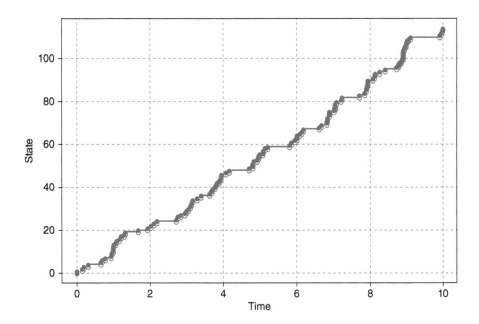

□

4.3   Applications of Nonhomogeneous Poisson Process

APPLICATION 4.1. In Application 3.1, we modeled the occurrence of earth-quakes in Southern California between 2012 and 2018 via a Poisson model and concluded that it fits the data well. Now we will try to fit a nonhomogeneous Poisson process to a larger data set covering the time span between 2010 and 2020. In this larger time interval, the event rate doesn't stay constant.

First, we estimate the intensity rate function. The code below calculates λ as the ratio of the number of earthquakes per month over the number of days in that month, producing the estimated daily rate for each of the 120 months. One outlying value is removed (for July of 2019), leaving us with 119 values. As the time variable, we compute the accrued number of days from 09/02/2010 to the median day for each of the 119 months. Next, we plot the estimated λ against time and fit a fourth-degree polynomial regression. We then define the intensity rate function $\widehat{\lambda}(t)$ as the fourth-degree polynomial with the estimated coefficients. After that, we subdivide the timeline between 09/02/2010 and 08/28/2020 into 9 bins of size 400 days and calculate the observed number of earthquakes in each bin. Then we compute the expected number of occurrences in each bin as the integral of $\widehat{\lambda}(t)$ between the lower and upper values of time in this bin. Finally, we compute the chi-squared statistic for the goodness-of-fit test and output the p-value.

The code and all the necessary outputs follow.

```
eq.data<- read.csv(file="./earthquakedata2010-2020.csv",
header=TRUE, sep=",")

#creating date-time variable
eq.data$datetime<- as.POSIXct(paste(as.Date(eq.data$DATE),
eq.data$TIME))

#computing lag
eq.data$datetime.lag<- c(0,head(eq.data$datetime, -1))
#removing first row
eq.data<-eq.data[-1,]

#computing interarrival times (in hours)
eq.data$elapsed.time<- (as.numeric(eq.data$datetime)
-as.numeric(eq.data$datetime.lag))/3600

#removing immediate aftershocks (within 1 hour)
eq.data<- eq.data[eq.data$elapsed.time>1,]

#creating year-month variable
eq.data$year.month<- format(as.Date(eq.data$DATE), "%Y-%m")

#creating unique year-month and number of earthquakes per
month
freq.month<- data.frame(table(eq.data$year.month))
year.month.unique<-freq.month[,1]
neq.month<- freq.month[,2]
neq.month
```

```
  [1]  37  43  29  35  29  19  19  26  18  23  24  11  12  16  19   6
 [17]  10  15  10  16  19  12  20  27  14  12   8  10   9   8   8  10
 [33]  20  14   7  14  14  12   7  12   8   8  17   9   7  11  21   9
 [49]  10  14   5   9   5  10   7   7  14   9   4   2   2   9   5  16
 [65]  10  13   7   8   4  13   6  10  17  11   3  10   6   5  10  11
 [81]   9  10  10   8   8   8   6   9   9   5   6  11   9   7   7  12
 [97]  11  10   9   7  14  11   3   9   5  18 126  35  20  11  25  21
[113]  15  12  12  16  19  30  10  15
```

```
#removing outlier (July, 2019, 107th entry)
year.month.unique<- year.month.unique[-107]
neq.month<- neq.month[-107]

#computing number of days per month
library(lubridate)
day1.month <- ymd(paste(year.month.unique,"01", sep="-") )
library(Hmisc)
ndays.month<- monthDays(as.Date(day1.month, "%Y-%m-%d"))

#estimating intensity rate of earthquakes per day
lambda<- neq.month/ndays.month

#computing cumulative number of days until median day of each
month
median.time<- c()
ndays.total<- c()
median.time[1]<- ndays.month[1]/2
ndays.total[1]<- ndays.month[1]

for (i in 2:length(ndays.month)) {
median.time[i]<- ndays.total[i-1] + ndays.month[i]/2
   ndays.total[i]<- ndays.total[i-1] + ndays.month[i]
}

#plotting lambda against median time
plot(median.time, lambda)
```

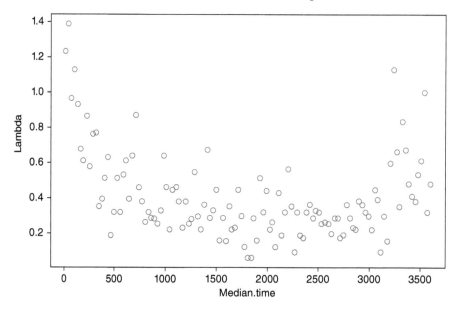

```
#regressing lambda on time
median.time.re<- median.time/1000
median.time.sq<- median.time.re^2
median.time.cu<- median.time.re^3
median.time.qd<- median.time.re^4
glm(lambda ~ median.time.re + median.time.sq + median.time.cu
+ median.time.qd)
```

```
Coefficients:
(Intercept)  median.time.re  median.time.sq  median.time.cu
   1.11674        -1.75263        1.41060        -0.49950

median.time.qd
      0.06504
```

```
#adding fitted line
lambda.fn<- function(t) { 1.11674-1.75263*(t/1000)+1.41060*
(t/1000)^2 -0.49950*(t/1000)^3+0.06504*(t/1000)^4 }

lines(median.time, lambda.fn(median.time), lwd=2,
col="blue")
```

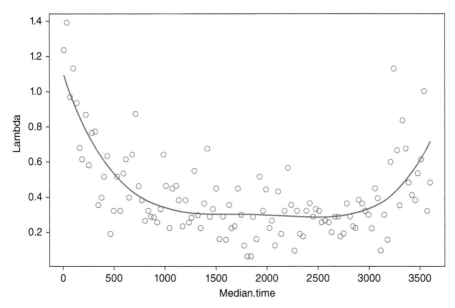

```
#conducting goodness-of-fit test

#binning times
time.binned<- as.factor(ifelse(as.Date(eq.data$DATE)
<"2011/10/07","1", ifelse(as.Date(eq.data$DATE)>="2011/10/07"&
as.Date(eq.data$DATE)
<"2012/11/10","2", ifelse(as.Date(eq.data$DATE)>="2012/11/10"
& as.Date(eq.data$DATE)<"2013/12/15","3",
ifelse(as.Date(eq.data$DATE)
>="2013/12/15"& as.Date(eq.data$DATE)<"2015/01/19","4",
ifelse(as.Date(eq.data$DATE)>="2015/01/19"&
as.Date(eq.data$DATE)
<"2016/02/23","5", ifelse(as.Date(eq.data$DATE)>="2016/02/23"&
as.Date(eq.data$DATE)<"2017/03/29","6",
ifelse(as.Date(eq.data$DATE)
>="2017/03/29"& as.Date(eq.data$DATE)<"2018/05/03","7",
ifelse(as.Date(eq.data$DATE)>="2018/05/03"&
as.Date(eq.data$DATE)
<"2019/06/07", "8", "9")))))))))

#computing observed frequencies
obs<- table((time.binned))
```

```
#computing expected frequencies
exp<- c()
exp[1]<- integrate(lambda.fn, 0, 400)$value
exp[2]<- integrate(lambda.fn, 400, 800)$value
exp[3]<- integrate(lambda.fn, 800, 1200)$value
exp[4]<- integrate(lambda.fn, 1200, 1600)$value
exp[5]<- integrate(lambda.fn, 1600, 2000)$value
exp[6]<- integrate(lambda.fn, 2000, 2400)$value
exp[7]<- integrate(lambda.fn, 2400, 2800)$value
exp[8]<- integrate(lambda.fn, 2800, 3200)$value
exp[9]<- sum(obs)-sum(exp)

obs
```

```
  1   2   3   4   5   6   7   8   9
326 198 139 141 108 112 113 121 376
```

```
round(exp,1)
```

333.5 192.9 137.7 123.2 120.7 117.3 116.2 136.7 355.7

```
#computing chi-squared statistic
print(chi.sq<- sum((obs-exp)^2/exp))
```

7.494979

```
#computing p-value
print(p.value<- 1-pchisq(chi.sq, df=3))
```

0.05768759

Note that we estimated five parameters in the polynomial regression, therefore, we needed to pick at least seven bins to have a non-degenerate number of degrees of freedom for the test. The total time span in the data set is $3,647$ days (without the outlier), so it was reasonable to divide the range into 9 bins of size 400 days each (the last bin has 447 days). Nine bins result in 3 degrees of freedom, $(df = 9 - 1 - 5 = 3)$.

Looking at the chi-squared statistic and the corresponding p-value, we can conclude at the 5% level of significance that the mean values in this process are well modeled by the fitted integrated intensity rate function.

It remains to show that the interarrival times have an exponential distribution. It is not possible to do it rigorously because of the nonhomogeneous nature of the process, but at least we can construct histograms for the interarrival times in each of the nine bins to see that they have the shape of an exponential density. As an illustration, below we present the histograms for the first and the last bins.

```
int1<- eq.data$elapsed.time[as.Date(eq.data$DATE)
<"2011/10/07"]
hist(int1, main="", xlab="", ylab="", col="light blue")
```

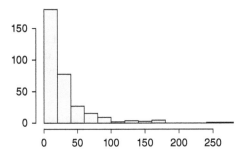

```
int9<- eq.data$elapsed.time[as.Date(eq.data$DATE)
>="2019/06/07"]
hist(int9, main="", xlab="", ylab="", col="light blue")
```

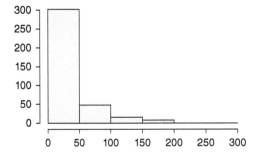

□

APPLICATION 4.2. Reliability engineers are concerned with the ability of manufactured systems or components to function without failure. Once failed, the item is repaired (in a repairable system) or replaced (in a non-repairable system). A stochastic model of the number of failures that has been widely used in practice by reliability engineers is a nonhomogeneous Poisson process with the *power-law intensity rate* (or *repair rate*) $\lambda(t) = \alpha\beta\, t^{\beta-1}$, $\alpha, \beta > 0$, $t \geq 0$. This function is very flexible because it models increasing rates if $\beta > 1$, or decreasing rates if $0 < \beta < 1$. If $\beta = 2$, the failure rate function degenerates into $\lambda(t) = 2\alpha\, t$, which corresponds to the homogeneous Poisson process.

(a) Let us study this model. Denote by $\{N(t), t \geq 0\}$ the number of failures by time t. The integrated intensity function is $\Lambda(t) = \int_0^t \alpha\beta\, u^{\beta-1}\, du = \alpha t^\beta$, $\alpha, \beta > 0, t \geq 0$. The probability mass function of $N(t)$ has the form

$$\mathbb{P}(N(t) - N(s) = n) = \frac{\left(\alpha t^\beta - \alpha s^\beta\right)^n}{n!}\, e^{-\alpha(t^\beta - s^\beta)}, \quad t \geq s \geq 0.$$

Note that the expected value of $N(t)$ is $\mathbb{E}(N(t)) = \Lambda(t) = \alpha t^\beta$.

An important question in reliability analysis is how to estimate α and β from the data. We address this question here. Two methods are commonly used to estimate the parameters. The first uses the linear regression approach, whereas the second one produces the maximum likelihood estimator.

METHOD 1 (LINEAR REGRESSION). Since $\mathbb{E}(N(t)) = \alpha t^\beta$, we can state the empirical analog $\widehat{N}(t) = \widehat{\alpha}\, t^{\widehat{\beta}}$, or, equivalently, $\ln(\widehat{N}(t)) = \ln\widehat{\alpha} + \widehat{\beta}\ln t$. Thus, $\ln(\widehat{N}(t))$ can be regressed linearly on $\ln t$ to obtain the estimated intercept $\ln\widehat{\alpha}$ and slope $\widehat{\beta}$.

To look at a numeric example, assume that S_k, $k = 1, 2, \ldots, 30$, the times to failure (in weeks) of certain auto parts during the pilot testing period, are as given in the table below. The second variable is $N(S_k) = k$, the total number of failures up to and including time S_k.

time	nfailures	time	nfailures	time	nfailures
2.36	1	10.16	11	19.70	21
2.96	2	10.87	12	20.99	22
4.71	3	11.32	13	21.56	23
5.23	4	13.36	14	22.57	24
6.16	5	14.52	15	22.79	25
7.15	6	16.19	16	24.02	26
7.33	7	16.84	17	25.8	27
8.20	8	17.19	18	26.49	28
8.45	9	18.18	19	27.13	29
9.31	10	18.72	20	28.05	30

Now we regress $\ln(N(S_k)) = \ln(k)$ on $\ln(S_k)$ to obtain $\widehat{\alpha} = e^{-0.6854} = 0.5039$, and $\widehat{\beta} = 1.2570$. The code and output are below.

```
reliability.data<- read.csv(file="./reliabilitydata.csv",
header=TRUE, sep=",")

x<- log(reliability.data$time)
y<- log(reliability.data$nfailures)

glm(y~x)
```

```
Coefficients:
(Intercept)              x
    -0.6854         1.2570
```

```
plot(x,y, xlab="ln(time)", ylab="ln(nfailures)")
lines(x, -0.6854+1.2570*x)
```

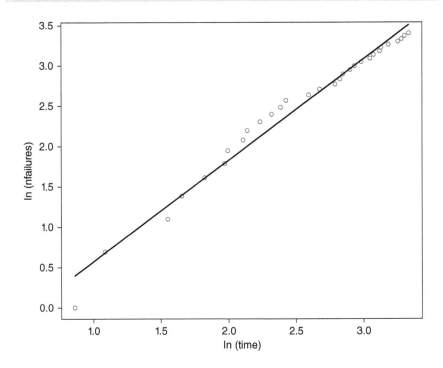

Using the known formulas for the estimators of the slope and intercept in linear regression, we write the explicit expressions for $\widehat{\beta}$ and $\widehat{\alpha}$. Here N is the total observed number of failures. In our example, $N = 30$. We have

$$\widehat{\beta} = \frac{\sum_{k=1}^{N} \left[\ln(k) \ln(S_k) \right] - \left[\ln(N!)/N \right] \left[\sum_{k=1}^{N} \ln(S_k) \right]}{\sum_{k=1}^{N} \left[\ln(S_k) \right]^2 - (1/N) \left[\sum_{k=1}^{N} \ln(S_k) \right]^2},$$

and

$$\widehat{\alpha} = \exp \left\{ (1/N) \left[\ln(N!) - \widehat{\beta} \sum_{k=1}^{N} \ln(S_k) \right] \right\}.$$

We can verify in R that these expressions give us the same results as above. Indeed,

```
print(beta1.hat<- (sum(x*y)-log(factorial(N))*mean(x))/
(sum(x^2) -N*(mean(x)) ^2))
```

1.256956

```
print(alpha1.hat<- exp(log(factorial(N))/N-beta1.hat*mean(x)))
```

0.5038849

Before we go to the second method of estimation, we need to define the distribution of the failure times $S_k, k = 1, 2, \ldots$. Given that the kth failure occurred at time s_k, the distribution of S_{k+1} is Weibull with the scale parameter α and the shape parameter β. The density is

$$f_{S_{k+1}|S_k}(t \mid s_k) = \alpha \beta t^{\beta-1} e^{-\alpha \left(t^\beta - s_k^\beta\right)}, \ t \geq s_k,$$

and the cumulative distribution function is

$$F_{S_{k+1}|S_k}(t \mid s_k) = 1 - e^{-\alpha \left(t^\beta - s_k^\beta\right)}, \ t \geq s_k.$$

METHOD 2 (MAXIMUM LIKELIHOOD ESTIMATOR). Suppose we observed N failures at times $S_1 = s_1, \ldots, S_N = s_N$. We write the likelihood function as

$$L(\alpha, \beta \mid s_1, \ldots, s_N) = f_{S_N|S_{N-1}}(s_N \mid s_{N-1}) f_{S_{N-1}|S_{N-2}}(s_{N-1} \mid s_{N-2}) \cdot \ \cdots \ \cdot$$

$$f_{S_2|S_1}(s_2 \mid s_1) f_{S_1}(s_1)$$

$$= \alpha \beta s_N^{\beta-1} e^{-\alpha \left(s_N^\beta - s_{N-1}^\beta\right)} \cdot \alpha \beta s_{N-1}^{\beta-1} e^{-\alpha \left(s_{N-1}^\beta - s_{N-2}^\beta\right)} \cdot \ \cdots \ \cdot$$

$$\alpha \beta s_2^{\beta-1} e^{-\alpha \left(s_2^\beta - s_1^\beta\right)} \cdot \alpha \beta s_1^{\beta-1} e^{-\alpha s_1^\beta}$$

$$= (\alpha \beta)^N \left(\prod_{k=1}^N s_k\right)^{\beta-1} e^{-\alpha s_N^\beta}.$$

Next, we find the expression for the log-likelihood function. We have

$$\ln(L) = N \ln \alpha + N \ln \beta + (\beta - 1) \sum_{k=1}^N \ln(s_k) - \alpha s_N^\beta.$$

Differentiating the log-likelihood function with respect to α and β and setting the expressions equal to 0, we obtain

$$\frac{\partial \ln(L)}{\partial \alpha} = \frac{N}{\alpha} - s_N^\beta = 0,$$

and

$$\frac{\partial \ln(L)}{\partial \beta} = \frac{N}{\beta} + \sum_{k=1}^{N} \ln(s_k) - \alpha\, s_N^\beta \ln(s_N) = 0.$$

From the first equation, $\alpha\, s_N^\beta = N$, and thus, the second equation can be rewritten as

$$\frac{N}{\beta} + \sum_{k=1}^{N} \ln(s_k) = \alpha\, s_N^\beta \ln(s_N) = N \ln(s_N).$$

From here,

$$\widehat{\beta} = \frac{N}{N \ln(s_N) - \sum_{k=1}^{N} \ln(s_k)} = \left[(1/N) \sum_{k=1}^{N} \ln(s_N/s_k) \right]^{-1}, \quad \text{and} \quad \widehat{\alpha} = N/s_N^{\widehat{\beta}}.$$

Going back to our numeric example, we write syntax in R, utilizing the variable x that contains the logs of failure times.

```
print(beta2.hat<- N/(N*x[N]-sum(x)))
```

1.236357

```
print(alpha2.hat<- N/exp(x[N]*beta2.hat))
```

0.4863609

(b) Another essential question in reliability analysis concerns the prediction of the next failure time. Suppose we have observed N failure times and the last one was at s_N. We can estimate S_{N+1} by its conditional mean value. We write

$$\mathbb{E}\big(S_{N+1} \,|\, S_N = s_N\big) = \int_{s_N}^{\infty} \alpha \beta\, t^\beta\, e^{-\alpha\left(t^\beta - s_N^\beta\right)}\, dt.$$

Now we use the substitution $u = \alpha t^\beta$, from where $\alpha\beta\, t^\beta\, dt = t\, du = \alpha^{-1/\beta}\, u^{1/\beta}\, du$. So, we obtain

$$\mathbb{E}\big(S_{N+1} \,|\, S_N = s_N\big) = \alpha^{-1/\beta}\, e^{\alpha\, s_N^\beta} \int_{\alpha\, s_N^\beta}^{\infty} u^{1/\beta}\, e^{-u}\, du.$$

The integral can be calculated in R as an upper incomplete gamma function $\int_x^\infty u^{a-1} e^{-u} du$ with the parameter $a = 1/\beta + 1$, and the lower limit of integration $x = \alpha\, s_N^\beta$. The following code computes the prediction. It uses the maximum likelihood estimators of α and β.

```
library(pracma)
alpha2.hat^(-1/beta2.hat)*exp(alpha2.hat*exp(x[N])
^beta2.hat)*
gammainc(alpha2.hat*exp(x[N])^beta2.hat, 1/beta2.hat+1)[2]
```

```
28.80161
```

Thus, given that the 30th failure was observed at 28.05 weeks, we predict that the 31st failure will occur at 28.8 weeks. ☐

Exercises

EXERCISE 4.1. While still under the manufacturer's warranty, calculators break down during the first three years with the rate of 3 per year. Between three and ten years, the rate increases linearly from 3 per year to 17 per year.
(a) What stochastic model can be used to model the number of broken calculators? Specify all parameters.
(b) Find the probability that 50 calculators break down between year 4 and year 8.
(c) Find the average number of calculators that will break down between year 2 and year 10.

EXERCISE 4.2. Occurrence of wildfires in a certain area during a 120-day fire season can be modeled as a nonhomogeneous Poisson process with the intensity function $\lambda(t) = -0.000025\, t^3 + 0.002\, t^2 + 0.12\, t$, where $0 \le t \le 120$.
(a) Plot the intensity function. Discuss its behavior. When is the peak of the intensity rate?
(b) Plot the integrated intensity function. Find the average number of wildfires per season.
(c) Find the average number of wildfires during the middle 50% of the season.

EXERCISE 4.3. Workers' injuries at an industrial manufacturing plant occur according to a nonhomogeneous Poisson process with the rate function $\lambda(t) = A/\sqrt{t}$, $t \ge 0$.

(a) Given that 30 injuries happened, on average, during the first year of plant's operation, find the value of A.

(b) Find the distribution of the times elapsing between two injuries. Simulate a trajectory of 100 injuries. What is the time range of the trajectory?

(c) Assuming that the 100th injury occurred 12 years and 3 months after the plant was opened, simulate a trajectory.

EXERCISE 4.4. Road traffic or airport noise data are often modeled over time as a nonhomogeneous Poisson process. It is assumed that noise pollution above a certain threshold has an intensity rate that allows cyclic behavior of observations. For instance, suppose an environmental noise pollution has the intensity rate $\lambda(t) = 10 * (1 + \cos(2\pi t))$, $t \geq 0$. Use the thinning method to simulate a trajectory on the interval of length 10.

EXERCISE 4.5. In the process of radioactive decay, photons are emitted according to a nonhomogeneous Poisson process with the intensity rate $\lambda(t) = 100e^{-0.5t}$, $t \geq 0$. Use each of the three simulation methods to generate a trajectory. Fix parameters as 20 events for the first method, and the length of the time interval as 0.25 in the other two methods.

EXERCISE 4.6. The National Weather Service website contains the data on fatal lightning strikes in the United States. The file *https://www.weather.gov/media/hazstat/80years.pdf* gives the counts of yearly lightning fatalities between 1940 and 2019.

(a) Plot the counts against year. Argue that the intensity rate function decays exponentially. Provide a possible explanation for this decay.

(b) Details of US lightning deaths (in particular, the dates) are provided on the same website *https://www.weather.gov/safety/lightning-victims*. The data are given for 2006-2019. Use the data to support the statement that this natural phenomenon is not governed by a nonhomogeneous Poisson process. Hint: Are the interarrival times exponentially distributed?

EXERCISE 4.7. The capacity of cargo container ships and port terminals are traditionally measured in twenty-foot equivalent units (TEUs), the number of 20-foot-long containers. Suppose that containership arrival to a port can be modeled by a nonhomogeneous Poisson process with the power-law intensity function $\lambda(t) = \alpha\beta\, t^{\beta-1}$, $\alpha, \beta > 0$, $t \geq 0$. The data for 27 arrivals (in units of 10,000 TEUs) are provided in the table below.

Arrivals	Days	Arrivals	Days	Arrivals	Days
1	3.72	10	21.16	19	41.23
2	5.45	11	23.33	20	43.03
3	8.65	12	25.26	21	45.43
4	10.33	13	26.77	22	48.13
5	12.54	14	30.19	23	49.82
6	14.83	15	32.74	24	52.27
7	15.82	16	35.75	25	53.32
8	18.04	17	37.51	26	55.91
9	19.05	18	38.85	27	59.10

(a) Estimate the parameters of the model using the regression approach.
(b) Estimate the parameters of the model using the maximum likelihood approach.
(c) Predict when the next 10,000 TEUs arrive at the port. Use both estimators from parts (a) and (b).

5

Compound Poisson Process

5.1 Definition of Compound Poisson Process

A stochastic process $\{X(t), t \geq 0\}$ is called a *compound Poisson process*[1] if $X(t) = \sum_{i=1}^{N(t)} Y_i$ where $\{N(t), t \geq 0\}$ is a Poisson process with rate λ, and Y_i's are independent and identically distributed random variables which are also independent of $\{N(t), t \geq 0\}$.

PROPOSITION 5.1. The mean and variance of a compound Poisson process are $\mathbb{E}(X(t)) = \lambda\, t\, \mathbb{E}(Y_1)$ and $\mathbb{V}ar(X(t)) = \lambda\, t\, \mathbb{E}(Y_1^2)$.

PROOF: The mean can be computed by conditioning on the value of $N(t)$. We write
$$\mathbb{E}(X(t)) = \mathbb{E}\big[\mathbb{E}(X(t) \mid N(t))\big] = \mathbb{E}\Big[\mathbb{E}\big(\sum_{i=1}^{N(t)} Y_i \mid N(t)\big)\Big]$$

$$= \mathbb{E}\big[N(t)\mathbb{E}(Y_1)\big] = \mathbb{E}(N(t))\,\mathbb{E}(Y_1) = \lambda\, t\, \mathbb{E}(Y_1).$$

Likewise, we compute the variance by conditioning on the value of $N(t)$. We get
$$\mathbb{V}ar(X(t)) = \mathbb{E}\big[\mathbb{V}ar(X(t) \mid N(t))\big] + \mathbb{V}ar\big[\mathbb{E}(X(t)|N(t))\big]$$

$$= \mathbb{E}\Big[\mathbb{V}ar\big(\sum_{i=1}^{N(t)} Y_i \mid N(t)\big)\Big] + \mathbb{V}ar\Big[\mathbb{E}\big(\sum_{i=1}^{N(t)} Y_i \mid N(t)\big)\Big] = \mathbb{E}\big(N(t)\mathbb{V}ar(Y_1)\big)$$

$$+ \mathbb{V}ar\big(N(t)\mathbb{E}(Y_1)\big) = \mathbb{E}(N(t))\mathbb{V}ar(Y_1) + \mathbb{V}ar(N(t))(\mathbb{E}(Y_1))^2 = \lambda\, t \mathbb{V}ar(Y_1)$$

$$+ \lambda\, t\, (\mathbb{E}(Y_1))^2 = \lambda\, t\big(\mathbb{E}(Y_1^2) - (\mathbb{E}(Y_1))^2\big) + \lambda\, t\, (\mathbb{E}(Y_1))^2 = \lambda\, t\, \mathbb{E}(Y_1^2). \quad \square$$

---

[1]The first treatment of the compound Poisson process is attributed to Filip Lundberg, a pioneer of the actuarial collective risk theory. In 1903, he wrote his Ph.D. dissertation at the University of Uppsala titled "Approximations of the Probability Function/Reinsurance of Collective Risks" (in Swedish).

DOI: 10.1201/9781003244288-5

EXAMPLE 5.1. Visitors walk into a casino in Las Vegas according to a Poisson process with a rate of 50 per hour. Ten percent of them will not gamble at all, others will lose independently a random number of dollars which we assume has a Uniform$(0, \$1,500)$ distribution. We need to model the casino's gain.

To this end, we focus only on the gamblers who are entering the casino. Their arrival can be modeled by a Poisson process $\{N(t),\ t \geq 0\}$ with rate $(0.9)(50) = 45$ per hour. Let Y_i denote the amount each gambler loses. We are given that Y_i's are independent and uniformly distributed with mean $\$1,500/2 = \$750$ and variance $(\$1,500)^2/12$. Consider $X(t) = \sum_{i=1}^{N(t)} Y_i$, the total sum of money that the gamblers lose at the casino within t hours. It is driven by a compound Poisson process.

(a) The casino's expected gain during a 12-hour period can be computed as
$\mathbb{E}(X(12)) = (45)(12)(\$750) = \$405,000.00$.

(b) The standard deviation of the gain is

$$\sqrt{\mathbb{V}ar(X(12))} = \sqrt{(45)(12)\big((\$1,500)^2/12 + (\$750)^2\big)} = \$20,124.61. \quad \square$$

EXAMPLE 5.2. Suppose the number of car accidents at a certain intersection can be modeled by a Poisson process with rate $\lambda = 3$ per month. Assume that the number of people who are involved in each accident is a binomially distributed random variable with parameters $n = 8$ and $p = 0.3$. Then the total number of people involved in car accidents on that intersection within a time period of t months can be modeled by a compound Poisson process $\{X(t) = \sum_{i=1}^{N(t)} Y_i,\ t \geq 0\}$ where $N(t) \sim Poisson(3t)$, and $Y_i,\ i = 1, 2, \dots, N(t)$, are independent random variables with $Bi(8, 0.3)$ distribution, also independent of $N(t)$. The average number of people involved in car accidents on this intersection within one year is $\mathbb{E}(X(12)) = \lambda t\, np = (3)(12)(8)(0.3) = 86.4$ with the standard deviation

$$\sqrt{\mathbb{V}ar(X(12))} = \sqrt{\lambda t\, \mathbb{E}(Y_1^2)} = \sqrt{\lambda t \big(\mathbb{V}ar(Y_1) + (\mathbb{E}(Y_1))^2\big)}$$

$$= \sqrt{\lambda t \,(np(1-p) + n^2 p^2)} = \sqrt{(3)(12)\big((8)(0.3)(0.7) + (8)^2(0.3)^2\big)} = 16.36582.$$

$\square$

EXAMPLE 5.3. Families enter a movie theater according to a Poisson process with rate $\tilde{\lambda} = 15$ per hour. The number of family members is distributed according to a zero-truncated Poisson distribution with rate $\lambda = 3$. We are interested in modeling the total number of moviegoers who enter the movie theater by time t.

Denote by $\{N(t), t \geq 0\}$ the Poisson process that describes family arrivals, and let Y_i, $i = 1, 2, \ldots, N(t)$, be the size of the ith family entering. We are given that $N(t) \sim Poisson(15t)$ and is independent of Y_i's, which, in turn, are independent and identically distributed with the probability mass function
$\mathbb{P}(Y_i = n) = \dfrac{\lambda^n}{n!} \dfrac{e^{-\lambda}}{1 - e^{-\lambda}}$, $n = 1, 2, \ldots$. The total number of movie goers can be described by a compound Poisson process $\{X(t) = \sum_{i=1}^{N(t)} Y_i, \ t \geq 0\}$. To find the mean and variance of this process, we first derive the expressions for the mean and second moment of a zero-truncated Poisson distribution. We write

$$\mathbb{E}(Y_1) = \frac{1}{e^\lambda - 1} \sum_{n=1}^{\infty} \frac{n\lambda^n}{n!} = \frac{\lambda}{e^\lambda - 1} \sum_{n=1}^{\infty} \frac{\lambda^{n-1}}{(n-1)!} = \frac{\lambda e^\lambda}{e^\lambda - 1},$$

and

$$\mathbb{E}(Y_1^2) = \frac{1}{e^\lambda - 1} \sum_{n=1}^{\infty} \frac{n^2 \lambda^n}{n!} = \frac{1}{e^\lambda - 1} \left[\sum_{n=1}^{\infty} \frac{n(n-1)\lambda^n}{n!} + \sum_{n=1}^{\infty} \frac{n\lambda^n}{n!} \right]$$

$$= \frac{1}{e^\lambda - 1} [\lambda^2 e^\lambda + \lambda e^\lambda] = \frac{\lambda(\lambda + 1) e^\lambda}{e^\lambda - 1}.$$

Thus, the average number of people who enter the movie theater during a t-hour period is

$$\mathbb{E}(X(t)) = \tilde{\lambda} t \, \mathbb{E}(Y_1) = \frac{\tilde{\lambda} t \lambda e^\lambda}{e^\lambda - 1},$$

with the standard deviation

$$\sqrt{\mathbb{V}ar(X(t))} = \sqrt{\tilde{\lambda} t \, \mathbb{E}(Y_1^2)} = \sqrt{\frac{\tilde{\lambda} t \lambda (\lambda + 1) e^\lambda}{e^\lambda - 1}}.$$

For instance, during a 6-hour period, the movie theater can expect $\mathbb{E}(X(6)) = \dfrac{(15)(6)(3) e^3}{e^3 - 1} = 284.1468$ visitors, with a standard deviation of $\sqrt{\mathbb{V}ar(X(6))} = \sqrt{\dfrac{(15)(6)(3)(3+1) e^3}{e^3 - 1}} = 33.71331$ visitors. $\quad\square$

5.2 Simulations in R

In this section, we use the setting of Example 5.1 and simulate trajectories of the casino's gain. First, we generate a Poisson process of gambler's arrivals and then generate the amounts lost by the gamblers, which are independent random variables uniformly distributed on $(0, \$1, 500)$. The sum of the loss amounts up to time t is the desired compound Poisson process. We generate the Poisson arrivals by the two methods described in Chapter 3, by fixing the number of arrivals at 20 gamblers and generating independent exponentially distributed interarrival times, and by fixing the time interval at 20 minutes and generating event times as the order statistics from the uniform distribution on $[0, 20]$. The codes and graphs are given below.

SIMULATION 5.1. (EXPONENTIAL INTERARRIVALS).

```
#specifying parameters
lambda<- 0.75
narrivals<- 20

#defining casino gain and time as vectors gain<- c()
time<- c()

#setting initial values
gain[1]<- 0
time[1]<- 0

#specifying seed
set.seed(50094)

#simulating trajectory
for (i in 2:(narrivals+1)) {
time[i]<- time[i-1] - 1/lambda*log(runif(1))
gain[i]<- gain[i-1] + runif(1,0, 1500)
}

#plotting trajectory
plot(time, gain, type="n", ylim=c(0,14000), xlab="Time
(min)", ylab="Casino gain ($)", panel.first = grid())

segments(time[-length(time)], gain[-length(time)],
time[-1]-0.15, gain[-length(time)], lwd=2, col="blue")

points(time, gain, pch=20, col="blue"))
points(time[-1], gain[-length(time)], pch=1, col="blue"))
```

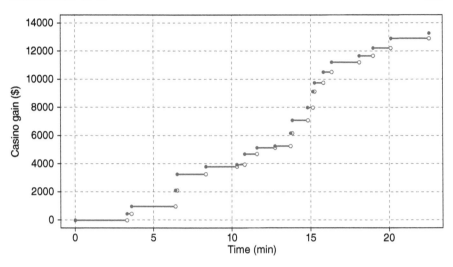

```
time[length(time)]
```

22.561

```
gain[length(gain)]
```

13291.65

In this simulated trajectory, 20 gamblers walked into the casino (a predetermined number of arrivals). They all came within 22.561 minutes and lost cumulatively \$13,291.65. □

SIMULATION 5.2. (UNIFORM ORDER STATISTICS).

```
#specifying parameters
t<- 20
lambda<- 0.75

#specifying seed
set.seed(41130)

#generating number of arrivals
narrivals<- rpois(1,lambda*t)

#defining vectors
gain<- c()
time<- c()
u<- c()
```

```
#setting initial values
gain[1]<- 0
u[1]<- 0

#generating standard uniforms and gain
for(i in 2:(narrivals+1)) {
u[i]<- runif(1)
gain[i]<- gain[i-1] + runif(1,0, 1500)
}

#computing event times
time<- t*sort(u)

#plotting trajectory
plot(time, gain, type="n", ylim=c(0,13000), xlab="Time
(min)", ylab="Casino gain ($)", panel.first = grid())

segments(time[-length(time)], gain[-length(time)],
time[-1]-0.15, gain[-length(time)], lwd=2, col="blue")

points(time, gain, pch=20, col="blue")
points(time[-1], gain[-length(time)], pch=1, col="blue")
```

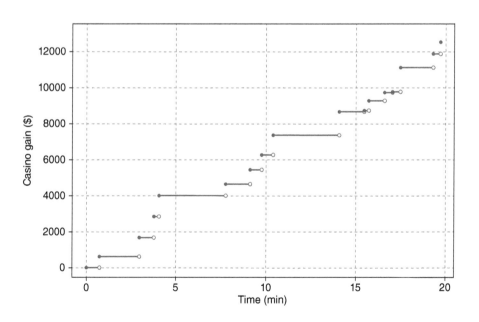

```
narrivals
```

16

```
gain[length(gain)]
```

12504.86

In this simulated trajectory, within a fixed time period of 20 minutes, 16 gamblers walked into the casino and lost a total of $12,504.86. □

5.3   Applications of Compound Poisson Process

APPLICATION 5.1. A compound Poisson *collective risk model* is a classical model in the actuarial field. It assumes that claims are submitted according to a Poisson distribution with rate λ, and that the amount of claims have a certain known distribution. Then the *aggregate claim amount* up to time t is a compound Poisson process $\left\{X(t) = \sum_{i=1}^{N(t)} Y_i,\ t \geq 0\right\}$ where $N(t)$ denotes the number (or *frequency*) of claims, and Y_i is the amount (or *severity*) of the ith claim. As in any compound Poisson model, $N(t)$ and all Y_i's are assumed independent. We know that $\mathbb{E}(X(t)) = \lambda\,t\,\mathbb{E}(Y_1)$ and $\mathbb{V}ar(X(t)) = \lambda\,t\,\mathbb{E}(Y_1^2)$.

The main question that actuaries have to answer is how much money should be collected in premiums so that the company will be able to pay the claims. Let $L(t) = X(t) - ct$ be the insurer's loss. It represents the difference between the total benefit payments that the company has to make and the amount of premiums collected over time with a constant rate c. It is customary to consider c of the form $c = (1 + \theta)\,\lambda\,\mathbb{E}(Y_1)$ where θ is termed a *security loading*. It means that up to time t, the company will collect in premiums the amount $ct = (1 + \theta)\,\lambda\,t\,\mathbb{E}(Y_1) = (1 + \theta)\,\mathbb{E}(X(t))$, which gives the company some cushion above the expected aggregate claim amount $\mathbb{E}(X(t))$ in case there are some unusually high claims. One of several ways to find the value of the security loading θ is to assume that the company wants to see a positive loss at most, say, 5% of the time. Thus, θ solves $\mathbb{P}(L(t) > 0) = 0.05$ or $\mathbb{P}\big(X(t) - (1 + \theta)\,\mathbb{E}(X(t)) > 0\big) = 0.05$. We can rewrite this identity as

$$\mathbb{P}\left(\frac{X(t) - \mathbb{E}(X(t))}{\sqrt{\mathbb{V}ar(X(t))}} > \theta\,\frac{\lambda\,t\,\mathbb{E}(Y_1)}{\sqrt{\lambda\,t\,\mathbb{E}(Y_1^2)}}\right) = 0.05.$$

Now, assuming that λ is large, we can use the Central Limit Theorem to conclude that $\dfrac{X(t) - \mathbb{E}(X(t))}{\sqrt{\mathbb{V}ar(X(t))}}$ has approximately a standard normal distribution. Hence, θ can be found as the solution of the equation

$$\mathbb{P}\left(Z > \theta \frac{\sqrt{\lambda t}\,\mathbb{E}(Y_1)}{\sqrt{\mathbb{E}(Y_1^2)}}\right) = 0.05.$$

That is,

$$\theta = 1.645 \frac{\sqrt{\mathbb{E}(Y_1^2)}}{\sqrt{\lambda t}\,\mathbb{E}(Y_1)}.$$

As an illustration, we consider data on storms downloaded from the National Oceanic and Atmospheric Administration's site (*https://www.ncdc.noaa.gov/stormevents/*). The data set contains dates, times, and amounts of damage (in units of $1,000) in all counties in Texas from March to April of 2020. The reported damage was done during a storm by hail, wind, tornado, flash flood, or lightning. There are a total of 85 rows in this data set. Damages range between $500 and $150,000, with two additional values of $500,000 and $800,000. Assuming that a single insurance company took care of all the claims, below we evaluate the security loading that this company must utilize when calculating premiums.

```
storm.data<- read.csv(file="./stormdata.csv", header=TRUE,
sep=",")

#creating date-time variable
storm.data$datetime<- as.POSIXct
(paste(as.Date(storm.data$Date), storm.data$Time))

#estimating event rate
print(nevents<- nrow(storm.data))
```

85

```
print(ndays<- (as.numeric(storm.data$datetime[nevents])
-as.numeric(storm.data$datetime[1]))/(24*3600))
```

46.09028

```
print(lambda.hat<- nevents/ndays)
```

1.844207

Within 46.09 days, there were 85 storms with tangible damage (resulting in the insurance claims). It means that the claims were submitted with a rate of 1.844207 per day. Finally, we estimate θ, using the empirical values of the first and second moments of the damage amounts.

```
#estimating security loading
print(theta<- 1.645*sqrt(mean(storm.data$Damage^2))/
    (sqrt(lambda.hat*ndays)*mean(storm.data$Damage)))
```

0.5516492

It means that the company should collect about 155% of the mean claim amount to hedge against large claims. □

Exercises

EXERCISE 5.1. The producer of a television game show with cash prizes wants to set a budget equal to the expected value plus one standard deviation of the aggregate cash prizes. The number of cash prizes given is a Poisson process with a rate of 1.5 per hour. Each episode lasts for 2 hours. The distribution of prize amounts is as follows:

Prize Amount	$5,000	$2,000	$500	$100
Probability	0.15	0.35	0.2	0.3

(a) Calculate the budget for 100 episodes of the game show.
(b) Simulate a trajectory of 100 games. For the trajectory, what is the amount of the budget left after the 100th game? If the amount is negative, during or after which episode did the producer run out of money?

EXERCISE 5.2. When a pharmacy bills the medical insurance company, the claims arrive as a Poisson process with the rate $\lambda = 60$ per day. The amounts of claims are independent and uniformly distributed between $30 and $300. It is also assumed that the amounts of the claims and the number of claims are independent.
(a) What is the expected aggregate claim amount that the medical insurance company receives within a 30-day period? What is the standard deviation of this amount?
(b) Use the Central Limit Theorem to approximate the probability that the aggregate claim amount will exceed $300,000 within a 30-day period.

EXERCISE 5.3. The photon detection process in X-ray computed tomography can be modeled as a compound Poisson process. The X-ray photons collide with a photon detector and then generate some number of light photons. The number of incident X-ray photons changes according to a Poisson process with a rate of λ per second. The number of light photons generated by each X-ray photon that is detected is a Poisson random variable with a rate of $\tilde{\lambda}$ per second.

(a) Define the aggregate number of light photons that are generated up to t seconds. Give the formula for the process, and the expressions for its mean and standard deviation.

(b) Assuming that the average rate of X-ray photons is 50 per second and the mean of light photons is 5, simulate 100 values of the aggregate number of light photons generated within a 10-second period.

(c) Construct a histogram for the 100 values generated in part (b). Is the histogram approximately bell-shaped? Explain.

EXERCISE 5.4. An insurance company receives claims according to a Poisson process $\{N(t),\ t \geq 0\}$ with rate λ. Assume that claim amounts X_i, $i = 1, 2, ...$, are independent and identically distributed, and are independent of claim arrival times S_i, $i = 1, 2, ...$ The present-day (day of policy issue) value of the amount of claim X_i made at time S_i is computed as $X_i e^{-\delta S_i}$, where δ is the force of interest. The present-day value of the total claim amount $P(t) = \sum_{i=1}^{N(t)} X_i e^{-\delta S_i}$ changes according to a compound Poisson process. Show that the mean of this process is $\mathbb{E}[P(t)] = \mathbb{E}(X_1)(\lambda/\delta)\left(1 - e^{-\delta t}\right)$.

EXERCISE 5.5. In radiobiology, when cells are exposed to radiation, DNA sometimes breaks, and the broken ends may abnormally rejoin resulting in chromosome aberrations. The number of ions that traverse through a cell nucleus by time t is modeled as a Poisson process with rate λ. Each traversal independently causes Y_i aberrations which have Poisson distribution with rate β. Let $X(t) = \sum_{i=1}^{N(t)} Y_i$ denote the total number of chromosomal aberrations by time t.

(a) Show that the compound Poisson process $X(t)$ has the Neyman type A distribution with the probability function given by

$$\mathbb{P}(X(t) = x) = \frac{\beta^x}{x!} e^{-\lambda t} \sum_{n=0}^{\infty} \frac{n^x (\lambda t)^n}{n!} e^{-\beta n}.$$

(b) Show that the mean of $X(t)$ is $\lambda t \beta$ and the variance is $\lambda t (\beta + \beta^2)$.

(c) For a fixed t, denote the sample mean of $X(t)$ by $\widehat{E}(X(t))$, sample variance by $\widehat{Var}(X(t))$, and empirical probability of zero $\mathbb{P}(X(t) = 0)$ by $\widehat{P}(0)$.

Show that β and λ may be estimated respectively by $\widehat{\beta} = \dfrac{\widehat{Var}(X(t))}{\widehat{E}(X(t))} - 1$ and

$\widehat{\lambda} = \dfrac{-\ln(\widehat{P}(0))}{t\left(1 - e^{-\widehat{\beta}}\right)}.$

EXERCISE 5.6. Cars arrive at a gas station according to a Poisson process. Each car driver buys gas independently of others for a dollar amount that has a gamma distribution. The spent amounts and the number of cars are independent. Data for times of car arrivals (in minutes) and dollar amounts spent are presented in the table below.

Arrival Time	Amount Spent, in $	Arrival Time	Amount Spent, in $	Arrival Time	Amount Spent, in $
0.15	23.67	28.81	69.67	47.94	20.38
3.81	25.55	32.36	25.39	49.73	34.95
5.67	38.54	32.76	30.86	50.72	29.23
6.61	31.31	32.92	50.53	50.86	36.51
13.14	74.20	33.22	24.93	51.99	37.77
13.57	32.78	33.51	27.49	52.36	34.41
15.68	29.70	34.40	22.56	52.89	23.35
22.83	35.83	35.76	21.38	53.64	32.95
23.35	22.17	39.08	45.53	55.03	21.27
23.77	34.96	41.03	39.14	55.29	37.32
23.77	24.20	42.05	26.02	56.82	20.30
24.69	26.01	42.38	21.35	63.02	32.59
26.94	24.07	45.66	33.88		

(a) Argue that the total dollar amount that the gas station receives can be modeled by a compound Poisson process. Write down the expression for the process and describe all parameters.

(b) Estimate the parameters of the model using the method of moments. Show that the estimators of α and β of the gamma distribution are $\widehat{\alpha} =$
$$\frac{n\overline{Y}^2}{\sum_{i=1}^{n} Y_i^2 - n\overline{Y}^2} \text{ and } \widehat{\beta} = \frac{\overline{Y}}{\widehat{\alpha}}.$$

(c) Plot histograms for the interarrival times and the amount spent, with fitted distribution curves.

(d) Write down the estimates of the mean and standard deviation of the total dollar amount at 1 hour. Use the parameter estimates obtained in part (b).

6

Conditional Poisson Process

6.1 Definition of Conditional Poisson Process

A counting process $\{N(t),\, t \geq 0\}$ is called a *conditional* (or *mixed*) Poisson process[1] if $N(t)$ has a Poisson distribution with rate Λ where Λ is a random variable with a known distribution $f_\Lambda(\lambda)$. The rate Λ is referred to as a *random intensity rate*.

The probability mass function for the conditional Poisson process is specified as a conditional probability

$$\mathbb{P}\big(N(t+s) - N(s) = n \,|\, \Lambda = \lambda\big) = \frac{(\lambda t)^n}{n!}\, e^{-\lambda t}, \quad t, s \geq 0,\ n = 0, 1, 2, \ldots.$$

The marginal distribution of $N(t)$ is found as

$$\mathbb{P}\big(N(t+s) - N(s) = n\big) = \int_0^\infty \frac{(\lambda t)^n}{n!}\, e^{-\lambda t}\, f_\Lambda(\lambda)\, d\lambda, \quad t, s \geq 0, n = 0, 1, 2, \ldots.$$

REMARK 6.1. A conditional Poisson process has stationary increments. It is reflected in the above formulas for conditional and marginal distributions of $N(t)$. The above expressions do not depend on s, the beginning of the interval, only on t, the length of the interval. The increments are not independent, though. It is easily seen if we write

$$\mathbb{P}\big(N(s) = n,\, N(t-s) = m\big) = \int_0^\infty \mathbb{P}\big(N(s) = n,\, N(t-s) = m \,|\, \Lambda = \lambda\big) f_\Lambda(\lambda)\, d\lambda$$

$$= \int_0^\infty \mathbb{P}\big(N(s) = n \,|\, \Lambda = \lambda\big) \mathbb{P}\big(N(t - s) = m \,|\, \Lambda = \lambda\big) f_\Lambda(\lambda)\, d\lambda$$

---

[1] Introduced in Dubourdieu, J. (1938). "Remarques relatives a la théorie mathématique de l'assurance-accidents." *Bulletin Trimestriel de l'Institut des Actuaires Français*, 49: 76 – 126.

$$\neq \int_0^\infty \mathbb{P}\big(N(s) = n \,|\, \Lambda = \lambda\big)\, f_\Lambda(\lambda)\, d\lambda \int_0^\infty \mathbb{P}\big(N(t-s) = m \,|\, \Lambda = \lambda\big)\, f_\Lambda(\lambda)\, d\lambda$$

$$= \mathbb{P}(N(s) = n)\, \mathbb{P}(N(t-s) = m). \quad \square$$

PROPOSITION 6.1. The mean of $N(t)$ is $\mathbb{E}(N(t)) = t\, \mathbb{E}(\Lambda)$, and the variance is $Var(N(t)) = t^2\, Var(\Lambda) + t\, \mathbb{E}(\Lambda)$.

PROOF: Conditioning on Λ, we write

$$\mathbb{E}(N(t)) = \mathbb{E}\big[\mathbb{E}(N(t)\,|\,\Lambda)\big] = \mathbb{E}(\Lambda\, t) = t\mathbb{E}(\Lambda),$$

and

$$Var(N(t)) = Var\big[\mathbb{E}(N(t)\,|\,\Lambda)\big] + \mathbb{E}\big[Var(N(t)\,|\,\Lambda)\big]$$

$$= Var\big[\Lambda\, t\big] + \mathbb{E}\big[\Lambda\, t\big] = t^2\, Var(\Lambda) + t\, \mathbb{E}(\Lambda). \quad \square$$

EXAMPLE 6.1. Assume that Λ in a conditional Poisson process has the probability mass function

$$\mathbb{P}(\Lambda = \lambda_0) = p_0 \ \text{ and } \ \mathbb{P}(\Lambda = \lambda_1) = 1 - p_0.$$

The marginal distribution of $N(t)$ is

$$\mathbb{P}(N(t) = n) = \frac{(\lambda_0\, t)^n}{n!}\, e^{-\lambda_0 t}\, p_0 + \frac{(\lambda_1\, t)^n}{n!}\, e^{-\lambda_1 t}(1 - p_0), n = 0, 1, 2, \dots.$$

This is a mixture of two Poisson processes with rates λ_0 and λ_1 and the mixture weights p_0 and $1 - p_0$. As a mixture of two Poisson processes, $N(t)$ has mean $\mathbb{E}(N(t)) = \lambda_0\, t\, p_0 + \lambda_1\, t\, (1 - p_0) = t\, (\lambda_0\, p_0 + \lambda_1(1 - p_0)) = t\, \mathbb{E}(\Lambda)$, and variance

$$Var(N(t)) = p_0(\lambda_0\, t + (\lambda_0\, t)^2) + (1 - p_0)(\lambda_1\, t + (\lambda_1\, t)^2) - \big(\lambda_0\, t\, p_0 + \lambda_1\, t\, (1 - p_0)\big)^2$$

$$= t^2\, (\lambda_0^2\, p_0 + \lambda_1^2\, (1 - p_0) - (\lambda_0\, p_0 + \lambda_1(1 - p_0))^2) + t\, (\lambda_0\, p_0 + \lambda_1(1 - p_0))$$

$$= t^2\, Var(\Lambda) + t\, \mathbb{E}(\Lambda). \quad \square$$

EXAMPLE 6.2. Suppose that in a conditional Poisson process, Λ has an exponential distribution with mean λ. The marginal probability mass function of $N(t)$ is

$$\mathbb{P}(N(t) = n) = \int_0^\infty \frac{(u\, t)^n}{n!}\, e^{-ut}\, \frac{1}{\lambda}\, e^{-u/\lambda}\, du$$

$$= \frac{t^n}{(t + 1/\lambda)^{n+1}\, \lambda} \int_0^\infty \frac{u^n\, (t + 1/\lambda)^{n+1}}{n!}\, e^{-(t+1/\lambda)u}\, du$$

$$= \frac{t^n}{(t + 1/\lambda)^{n+1}\, \lambda} = \frac{1}{\lambda t + 1}\left(1 - \frac{1}{\lambda t + 1}\right)^n, \quad n = 0, 1, 2, \dots.$$

Therefore, the marginal distribution of $N(t)$ is geometric. It represents the number of failures until the first success. The success probability is $p = 1/(\lambda t + 1)$. It is known that the mean is $\dfrac{1 - p}{p} = \dfrac{1 - 1/(\lambda t + 1)}{1/(\lambda t + 1)} = \lambda t = t\,\mathbb{E}(\Lambda)$, and variance is $\dfrac{1 - p}{p^2} = \dfrac{1 - 1/(\lambda t + 1)}{1/(\lambda t + 1)^2} = \lambda t (\lambda t + 1) = t^2 \lambda^2 + t\lambda = t^2\,\mathbb{V}ar(\Lambda) + t\,\mathbb{E}(\Lambda)$. □

6.2   Simulations in R

SIMULATION 6.1. (EXPONENTIAL INTERARRIVALS). Consider the setting of Example 6.1. Suppose $p_0 = 0.3, \lambda_0 = 4$, and $\lambda_1 = 0.5$. Below we simulate 5 trajectories with 20 Poisson occurrences each, by generating exponential interarrival times. We first choose the rate randomly from the given binary distribution. Three of the trajectories have a rate of 4 and the other two have a rate of 0.5.

```
#specifying parameters
p<- c(0.3, 0.7)
lambda<- c(4, 0.5)
njumps<- 20

#specifying states and times as data frames
time<- data.frame()
N<- data.frame()

#specifying seed
set.seed(6335044)

#simulating trajectories
for(j in 1:5) {

#fixing the value for rate
Lambda<- lambda[sample(1:2, 1, prob=p)]

#setting initial values
time[1,j]<- 0
N[1,j]<- 0

#simulating trajectory
i<- 2
```

```
repeat {
time[i,j]<- time[i-1,j]+round((-1/Lambda)*log(1-runif(1)),3)
-0.001
N[i,j]<- N[i-1,j]

if(i==2*njumps+2) break
else {
time[i+1,j]<- time[i,j]+0.001
  N[i+1,j]<- N[i,j]+1
    i<- i+2
      }
 }
}


#plotting trajectories
matplot(time, N, type="l", lty=1, lwd=2, col=c("blue",
"green", "red", "purple", "orange"), xlab="Time",
ylab="State", panel.first=grid())
```

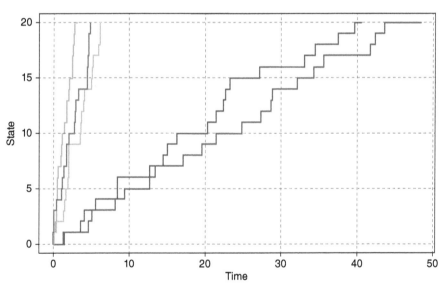

□

SIMULATION 6.2. (UNIFORM ORDER STATISTICS). Now we use the uniform order statistics method described in earlier chapters to simulate five trajectories. We fix the end-time (say, $t = 30$), and then pick the rate λ from the binary distribution. After that, we randomly choose the number of events from the Poisson (λt) distribution. Note that we expect trajectories with $\lambda = 0.5$ to have about $(30)(0.5) = 15$ occurrences, whereas trajectories with $\lambda = 4$ will have about $(30)(4) = 120$ occurrences. Below is the code and the graph. Two trajectories happened to have a rate of 4 and the other three have a rate of 0.5.

```
#specifying parameters
t<- 30
p<- c(0.3, 0.7)
lambda<- c(4, 0.5)

#specifying states and times as data frames
time<- data.frame()
N<- data.frame()

#specifying seed
set.seed(1902238)

#simulating trajectories
for(j in 1:5) {

#fixing the value for rate
Lambda<- lambda[sample(1:2, 1, prob=p)]

#setting initial values
time[1,j]<- 0
N[1,j]<- 0

#generating total number of jumps
njumps<- rpois(1,Lambda*t)

#generating standard uniforms
u<- c()
u[1]<- 0
for(i in 2:njumps)
u[i]<- runif(1)

#computing event times
s<- t*sort(u)
```

```
#generating jumps
for (i in seq(2, 2*njumps, 2)) {
time[i,j]<- s[i/2]-0.001
  time[i+1,j]<- s[i/2]
    N[i,j]<- N[i-1,j]
      N[i+1,j]<- N[i-1,j]+1
        }
}
#plotting trajectories
matplot(time, N, type="l", lty=1, lwd=2, col=c("blue",
"green", "red", "purple", "orange"), xlab="Time",
ylab="State", panel.first=grid())
```

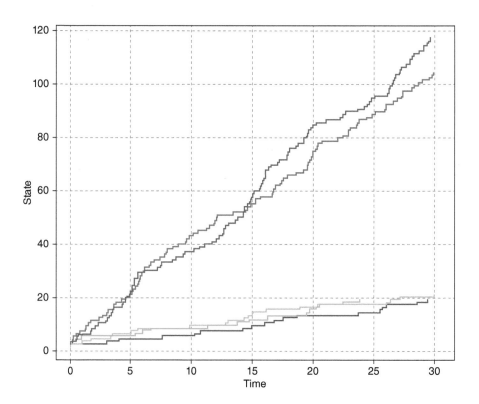

□

6.3 Applications of Conditional Poisson Process

APPLICATION 6.1. Agronomists model the presence of Colorado potato bee-tles as a conditional Poisson process. Suppose a plot of fertile soil is planted with potatoes. The width of the plot is fixed, so the area is proportional to the length. The number of egg clusters Λ that are located within a stretch of the plot of a certain length has a Poisson distribution with mean λ_0. Each cluster contains between 10 and 30 eggs, but only several of them hatch. Suppose the number of eggs within a cluster that hatch follows a Poisson distribution with a rate r. Combining the results, we obtain that the total number of hatched eggs $N(\ell)$ on a stretch of the plot of length ℓ has a Poisson distribution with the rate $r\,\Lambda$, where $\Lambda \sim Poisson(\lambda_0)$.

(a) The expected value of $N(\ell)$ is computed by conditioning on Λ. We write $\mathbb{E}(N(\ell)) = \mathbb{E}\big[\mathbb{E}(N(\ell)\,|\,\Lambda)\big] = \mathbb{E}(r\ell\Lambda) = r\lambda_0\ell$. Likewise, the variance is computed by conditioning on Λ. We have $\mathbb{V}ar(N(\ell)) = \mathbb{V}ar\big[\mathbb{E}(N(\ell)\,|\,\Lambda)\big] + \mathbb{E}\big[\mathbb{V}ar(N(\ell)\,|\,\Lambda)\big] = \mathbb{V}ar(r\ell\Lambda) + \mathbb{E}(r\ell\Lambda) = (r\ell)^2\lambda_0 + (r\ell)\lambda_0 = \lambda_0(r\ell)(r\ell+1)$.

To illustrate these with a numeric example, assume $r = 5, \ell = 2$, and $\lambda_0 = 3$. We evaluate $\mathbb{E}(N(\ell)) = (5)(3)(2) = 30$ and $\mathbb{V}ar(N(\ell)) = (3)(5)(2)((5)(2) + 1) = 330$.

(b) The marginal distribution of $N(\ell)$ is

$$\mathbb{P}(N(\ell) = n) = \mathbb{E}\big[\mathbb{P}(N(\ell) = n\,|\,\Lambda)\big] = \mathbb{E}\Big[\frac{(r\Lambda\,\ell)^n}{n!}\,e^{-r\Lambda\,\ell}\Big]$$

$$= \frac{(r\ell)^n}{n!}\,\mathbb{E}\big[\Lambda^n\,e^{-r\ell\Lambda}\big] = \frac{(r\ell)^n}{n!}\,M_\Lambda^{(n)}(-r\ell), \quad n = 0, 1, 2, \ldots,$$

where $M_\Lambda^{(n)}(-r\ell)$ is the nth derivative of the moment generating function of Λ, $M_\Lambda(t) = \exp\{\lambda_0\,(e^t - 1)\}$, computed at $t = -r\ell$. These derivatives have to be computed numerically. Below is the R code that calculates probabilities for $n = 0, \ldots, 10$.

```
#specifying parameters
r<- 5
l<- 2
t<- -r*l
lambda0<- 3
```

```
#computing probabilities
prob<- c()
M<- expression(exp(lambda0*(exp(t)-1)))
prob[1]<- eval(M)

for (n in 2:11) {
M<- D(M,"t")
prob[n]<- (r*l)^n/factorial(n)*eval(M)
 }

prob
```

```
[1]  0.0497938498 0.0003390956 0.0011304726
[4]  0.0028269513 0.0056569821 0.0094385700
[7]  0.0135130114 0.0169646392 0.0190127388
[10] 0.0193392713 0.0181755410
```

□

APPLICATION 6.2. An auto insurance company models the number of claims up to time t, $N(t)$, as a Poisson process with random intensity rate Λ which represents an accident-proneness index of a policyholder. The values of Λ are distributed according to a gamma distribution with mean α/β and variance α/β^2.

(a) The expectation and variance of $N(t)$ are $\mathbb{E}(N(t)) = t\mathbb{E}(\Lambda) = \alpha t/\beta$, and $\mathbb{V}ar(N(t)) = t^2\mathbb{V}ar(\Lambda) + t\mathbb{E}(\Lambda) = \alpha t^2/\beta^2 + \alpha t/\beta$.

To compute these quantities for some specific values of the parameters, let's assume that $\alpha = 0.3$ and $\beta = 1$. The average total number of claims per policyholder that the company has to deal with every 5 years is $\mathbb{E}(N(5)) = (0.3)(5)/(1) = 1.5$. The standard deviation is

$$\sqrt{\mathbb{V}ar(N(5))} = \sqrt{(0.3)(5)^2/(1)^2 + (0.3)(5)/(1)} = 3.$$

(b) The marginal probability mass function of $N(t)$ has the form

$$\mathbb{P}(N(t) = n) = \int_0^\infty \frac{(\lambda t)^n}{n!} e^{-\lambda t} \frac{\lambda^{\alpha-1}\beta^\alpha}{\Gamma(\alpha)} e^{-\beta\lambda} \, d\lambda$$

$$= \frac{t^n}{n!} \frac{\Gamma(n+\alpha)}{\Gamma(\alpha)} \frac{\beta^\alpha}{(t+\beta)^{n+\alpha}} \int_0^\infty \frac{\lambda^{n+\alpha-1}(t+\beta)^{n+\alpha}}{\Gamma(n+\alpha)} e^{-(t+\beta)\lambda} \, d\lambda$$

$$= \frac{(n+\alpha-1)!}{n!\,(\alpha-1)!} \left(\frac{\beta}{t+\beta}\right)^\alpha \left(1 - \frac{\beta}{t+\beta}\right)^n.$$

Therefore, $N(t)$ has a *gamma–Poisson mixture distribution*, which has the algebraic form of a negative binomial distribution with parameters α and $p = \dfrac{\beta}{t+\beta}$. Since α assumes real values that are not restricted to integers, this is not a true negative binomial distribution. The mean and variance of this distribution are

$$\mathbb{E}(N(t)) = \frac{\alpha(1-p)}{p} = \frac{\alpha t/(t+\beta)}{\beta/(t+\beta)} = \alpha t/\beta,$$

and

$$\mathrm{Var}(N(t)) = \frac{\alpha(1-p)}{p^2} = \frac{\alpha t/(t+\beta)}{\beta^2/(t+\beta)^2} = \frac{\alpha t(t+\beta)}{\beta^2} = \alpha t^2/\beta^2 + \alpha t/\beta.$$

These coincide with the expressions derived in part (a).

(c) We can also compute the conditional probability that Λ, the accident-proneness index of a policyholder, doesn't exceed some particular value λ, provided that the policyholder has made n claims within t years. We derive

$$\mathbb{P}(\Lambda \le \lambda \,|\, N(t) = n) = \frac{\mathbb{P}(N(t)=n, \Lambda \le \lambda)}{\mathbb{P}(N(t)=n)} = \frac{\int_0^\lambda \frac{(ut)^n}{n!} e^{-ut} \frac{u^{\alpha-1}\beta^\alpha}{\Gamma(\alpha)} e^{-\beta u}\, du}{\int_0^\infty \frac{(ut)^n}{n!} e^{-ut} \frac{u^{\alpha-1}\beta^\alpha}{\Gamma(\alpha)} e^{-\beta u}\, du}$$

$$= \frac{\int_0^\lambda u^{n+\alpha-1} e^{-(t+\beta)u}\, du}{\int_0^\infty u^{n+\alpha-1} e^{-(t+\beta)u}\, du} = \int_0^\lambda \frac{u^{n+\alpha-1}(t+\beta)^{n+\alpha}}{\Gamma(n+\alpha)} e^{-(t+\beta)u}\, du.$$

This is a gamma distribution with parameters $n+\alpha$ and $t+\beta$. The expected value of Λ for a policyholder who is known to have made n claims within t years is $\mathbb{E}(\Lambda \,|\, N(t)=n) = \frac{n+\alpha}{t+\beta}$.

For $\alpha = 0.3$ and $\beta = 1$, the expected accident-proneness index for a policyholder who has made two claims in 5 years is $\mathbb{E}(\Lambda \,|\, N(5)=2) = \frac{2+0.3}{5+1} = 0.3833$. □

Exercises

EXERCISE 6.1. Let $\{N(t),\ t \ge 0\}$ be a conditional Poisson process with the random intensity rate Λ. Show that for any $t \ge s \ge 0$,
(a) $\mathbb{C}ov\big(N(s),\ N(t) - N(s)\big) = s\,(t-s)\,\mathbb{V}ar(\Lambda)$.
(b) $\mathbb{C}ov\big(N(s),\ N(t)\big) = s\,t\,\mathbb{V}ar(\Lambda) + s\,\mathbb{E}(\Lambda)$.

EXERCISE 6.2. Suppose $\{N(t),\ t \geq 0\}$ is a conditional Poisson process and the random intensity rate Λ has density $f_\Lambda(\lambda)$, $\lambda > 0$. Show that

(a) The conditional cumulative distribution function of Λ, given $N(t) = n$, is

$$F_{\Lambda|N(t)}(\lambda|n) = \mathbb{P}\big(\Lambda \leq \lambda \mid N(t) = n\big) = \frac{\int_0^\lambda u^n\, e^{-ut}\, f_\Lambda(u)\, du}{\int_0^\infty u^n\, e^{-ut}\, f_\Lambda(u)\, du}.$$

(b) The conditional density function of Λ, given $N(t) = n$, is

$$f_{\Lambda|N(t)}(\lambda|n) = \frac{\lambda^n\, e^{-\lambda t}\, f_\Lambda(\lambda)}{\int_0^\infty \lambda^n\, e^{-\lambda t}\, f_\Lambda(\lambda)\, d\lambda}.$$

(c) The conditional expected value of Λ, given $N(t) = n$, is

$$\mathbb{E}\big[\Lambda \mid N(t) = n\big] = \frac{\int_0^\infty \lambda^{n+1}\, e^{-\lambda t}\, f_\Lambda(\lambda)\, d\lambda}{\int_0^\infty \lambda^n\, e^{-\lambda t}\, f_\Lambda(\lambda)\, d\lambda}.$$

EXERCISE 6.3. Suppose that 46% of all visitors of an amusement park are teenagers, 24% are adults, and the rest are kids. Assume that the number of visitors who come into the park during a busy hour can be modeled as a Poisson process with the random intensity rate that varies depending on age group: 4 per minute for teens, 2 per minute for adults, and 3 per minute for kids.

(a) Write down the model and specify all parameters. Find the mean and variance of the number of visitors within time t.

(b) Simulate five trajectories of the process with 200 visitors each.

(c) Simulate five trajectories of the process that depict arrivals within 1 hour.

EXERCISE 6.4. A credit union assigns to all its clients a rating value Λ in such a way that a client defaults on a credit account according to a Poisson process with rate Λ (per year). The distribution of Λ is uniform on $[0, 2]$.

(a) Find the average number of defaults that a client has within a 5-year period.

(b) Find the variance of the number of defaults within a 5-year period.

(c) Find the covariance between the number of defaults during the first 3 years and that during the subsequent 2 years. Hint: See Exercise 6.1(a).

(d) Find the covariance between the number of defaults during the first 3 years and that during the first 5 years. Hint: See Exercise 6.1(b).

(e) Find the probability that the client's rating value is less than 0.5, given that he has had two defaults within a 5-year period. Hint: See Exercise 6.2.

EXERCISE 6.5. Snow Water Equivalent (SWE) describes the amount of water contained within a snowpack, measured in inches. It is an important notion in environmental science, agriculture, and forestry. An annual SWE is well modeled by a conditional Poisson process. The amount of SWE is a Poisson process with rate Λ inches per year that is itself a Poisson random variable with a rate of 24.3.

(a) Compute the average and standard deviation of SWE for a 1-year period. For a 5-year period.

(b) Simulate five trajectories that reach 140 inches of SWE each.

(c) Simulate five trajectories spanning over a 7-year period.

EXERCISE 6.6. In the textile industry, a series issue is the number of defects in fabric per linear footage. Suppose quality control engineers model the number of defects as a conditional Poisson process with a random intensity rate Λ per yard for the fabric of standard width. The random variable Λ has a gamma distribution with a mean of 0.07 and a standard deviation of 0.01.

(a) Compute the expected number of defects in a 40-yard roll of fabric. Find the standard deviation of the number of defects in the roll.

(b) Given that four defects were found in a 40-yard roll of fabric, find the probability that Λ exceeds 0.08.

7

Birth-and-Death Process

7.1   Definition of Birth-and-Death Process

A continuous-time Markov chain $\{X(t),\ t \geq 0\}$ with state space $S = \{0, 1, 2, \ldots\}$ is called a *birth-and-death process*[1] if, given that the chain is in state n, the time to transition to state $n + 1$ is exponentially distributed with mean $1/\lambda_n$, and the time to transition to state $n - 1$ is exponentially distributed with mean $1/\mu_n$. The two waiting times are independent.

PROPOSITION 7.1. In a birth-and-death process, the transition probabilities are $P_{0,1} = 1$, $P_{n,n+1} = \frac{\lambda_n}{\lambda_n + \mu_n}$, and $P_{n,n-1} = \frac{\mu_n}{\lambda_n + \mu_n}$. All the other transition probabilities are 0's. The one-step transition probability matrix has the form

$$
\mathbf{P} = \begin{array}{c} \\ 0 \\ 1 \\ 2 \\ 3 \\ \cdots \end{array} \overset{\displaystyle \begin{array}{ccccc} 0 & 1 & 2 & 3 & \cdots \end{array}}{\left[\begin{array}{ccccc} 0 & 1 & 0 & 0 & \cdots \\ \frac{\mu_1}{\lambda_1 + \mu_1} & 0 & \frac{\lambda_1}{\lambda_1 + \mu_1} & 0 & \cdots \\ 0 & \frac{\mu_2}{\lambda_2 + \mu_2} & 0 & \frac{\lambda_2}{\lambda_2 + \mu_2} & \cdots \\ 0 & 0 & \frac{\mu_3}{\lambda_3 + \mu_3} & 0 & \cdots \\ \cdots & \cdots & \cdots & \cdots & \cdots \end{array} \right]}.
$$

PROOF: Consider two independent exponential random variables T_B and T_D with means $1/\lambda$ and $1/\mu$, respectively. The variable T_B represents the waiting time until a "birth," and T_D represents the waiting time until a "death." A "birth" occurs before a "death," if $T_B < T_D$. The probability of this event is

$$
\mathbb{P}(T_B < T_D) = \int_0^\infty \int_x^\infty \lambda e^{-\lambda x} \mu e^{-\mu y} \, dy \, dx = \int_0^\infty \lambda e^{-\lambda x} e^{-\mu x} \, dx = \frac{\lambda}{\lambda + \mu}.
$$

---

[1] The first example of a birth-and-death process was described in 1939 by William Feller, a renown Croatian-American mathematician, in "Die Grundlagen der Volterraschen Theorie des Kampfes ums Dasein in wahrscheinlichkeitstheoretischer Behandlung." *Acta Biotheoretica*, 5: 11 – 40.

DOI: 10.1201/9781003244288-7

Analogously, a "death" occurs before a "birth" if $T_D < T_B$, which happens with the complementary probability

$$\mathbb{P}(T_D < T_B) = 1 - \mathbb{P}(T_B < T_D) = 1 - \frac{\lambda}{\lambda + \mu} = \frac{\mu}{\lambda + \mu}. \quad \square$$

PROPOSITION 7.2. Suppose a birth-and-death process is in state n. Then the waiting time until a transition occurs is exponential with mean $1/(\lambda_n + \mu_n)$.

PROOF: Referring to the proof of the previous proposition, we see that a transition occurs at time T_B or time T_D, whichever happens first. That is, we need to find the distribution of $\min(T_B, T_D)$. We write

$$F_{\min(T_B, T_D)}(t) = \mathbb{P}(T_B \le t,\, T_B < T_D) + \mathbb{P}(T_D \le t,\, T_D < T_B)$$

$$= \int_0^t \int_x^\infty \lambda e^{-\lambda x} \mu e^{-\mu y}\, dy\, dx + \int_0^t \int_y^\infty \lambda e^{-\lambda x} \mu e^{-\mu y}\, dx\, dy$$

$$= \int_0^t \lambda e^{-\lambda x} e^{-\mu x}\, dx + \int_0^t e^{-\lambda y} \mu e^{-\mu y}\, dy$$

$$= \int_0^t (\lambda + \mu) e^{-(\lambda + \mu) u}\, du = 1 - e^{-(\lambda + \mu) t},$$

which is an exponential distribution with mean $1/(\lambda + \mu)$. $\square$

PROPOSITION 7.3. Consider a birth-and-death process, and denote by $P_n(t)$ the probability that at time t the process is in state n. The probabilities $P_n(t)$, $n = 0, 1, \ldots$, satisfy the system of the *Kolmogorov forward equations*:

$$\begin{cases} P_0'(t) = -\lambda_0\, P_0(t) + \mu_1\, P_1(t), \\ P_n'(t) = \lambda_{n-1}\, P_{n-1}(t) + \mu_{n+1}\, P_{n+1}(t) - (\lambda_n + \mu_n)\, P_n(t), \quad n = 1, 2, \ldots, \end{cases}$$

$$(7.1)$$

with the boundary condition $P_{n_0}(0) = 1$.

PROOF: We will omit the rigorous derivation of these equations but will explain their simple meaning. How can the process transition into state n? Only if there are $n-1$ particles and one more is born, or there are $n+1$ particles and one dies. That is, $\lambda_{n-1}\, P_{n-1}(t) + \mu_{n+1}\, P_{n+1}(t)$ represents the rate of change with respect to time of probability to transition into state n. By the same token, $(\lambda_n + \mu_n)\, P_n(t)$ gives the rate of change over time of probability to transition out of state n (one particle is born or dies). Thus, the expression on the right is the difference between the in and out rates, and so is the meaning of the derivative $P_n'(t)$ on the left. $\square$

REMARK 7.1. The mean and variance of $\{X(t),\ t \geq 0\}$, a birth-and-death process, are computed as

$$\mathbb{E}(X(t)) = \sum_{n=0}^{\infty} n\, P_n(t),\ \ \text{and}\ \ \mathbb{V}ar(X(t)) = \sum_{n=0}^{\infty} n^2\, P_n(t) - \big[\mathbb{E}(X(t))\big]^2.\ \ \square$$

EXAMPLE 7.1. A Poisson process is an example of a birth-and-death process with $\lambda_n = \lambda$ and $\mu_n = 0$, for $n = 0, 1, \ldots$. Because there are no "deaths," it is a *pure birth process*.

(a) We know that in this process, times until "births" are independent and exponentially distributed with mean $1/\lambda_n = 1/\lambda$.

(b) The transition probabilities are $P_{0,1} = 1$, $P_{n,n+1} = \lambda_n/(\lambda_n + \mu_n) = \lambda/(\lambda + 0) = 1$, and $P_{n,n-1} = \mu_n/(\lambda_n + \mu_n) = 0/(\lambda + 0) = 0$, which we know is true since only jumps of size $+1$ are admissible in a Poisson process.

(c) The probabilities $P_n(t)$, $n = 0, 1, \ldots$, satisfy the Kolmogorov forward equations $P_0'(t) = -\lambda\, P_0(t)$ and $P_n'(t) = \lambda\, P_{n-1}(t) - \lambda\, P_n(t)$, for $n = 1, 2, \ldots$, with the boundary condition $P_0(0) = 1$. The solution to these equations is the Poisson probability mass function $P_n(t) = \frac{(\lambda t)^n}{n!}\, e^{-\lambda t}$, $n = 0, 1, \ldots$, which can be verified by writing

$$P_0'(t) = \big(e^{-\lambda t}\big)' = -\lambda\, e^{-\lambda t} = -\lambda\, P_0(t),$$

and for $n \geq 1$,

$$P_n'(t) = \left(\frac{(\lambda t)^n}{n!}\, e^{-\lambda t}\right)' = \frac{\lambda^n}{n!}\left(n\, t^{n-1} - \lambda t^n\right) e^{-\lambda t} = \lambda \frac{(\lambda t)^{n-1}}{(n-1)!}\, e^{-\lambda t}$$

$$- \lambda \frac{(\lambda t)^n}{n!}\, e^{-\lambda t} = \lambda\, P_{n-1}(t) - \lambda\, P_n(t),\ \ \text{and}\ \ P_0(0) = \frac{0^0}{0!}\, e^0 = 1.\ \ \square$$

(d) Both the mean and variance of the Poisson process, as we know, are equal to λt.

EXAMPLE 7.2. A *linear birth-and-death process* $\{X(t),\ t \geq 0\}$ is a birth-and-death process with the parameters $\lambda_n = n\lambda$ and $\mu_n = \mu n$, $n = 0, 1, \ldots$. Note that this process state 0 is an absorbing state. We assume that the process starts in state 1.

(a) In this process, times until "births" and "deaths" are independent and exponentially distributed with mean $1/\lambda_n = 1/(n\lambda)$ and $1/\mu_n = 1/(n\mu)$, respectively. Time until a transition ("birth" or "death") are independent exponentially distributed random variables with mean $1/(\lambda_n + \mu_n) = 1/[n(\lambda+\mu)]$.

(b) The transition probabilities are $P_{0,1} = 0$, $P_{n,n+1} = \lambda_n/(\lambda_n + \mu_n) = \lambda/(\lambda + \mu)$, and $P_{n,n-1} = \mu_n/(\lambda_n + \mu_n) = \mu/(\lambda + \mu)$.

(c) The Kolmogorov forward equations in this case assume the form: $P_0'(t) = \mu P_1(t)$ and for $n = 1, 2, \ldots$, $P_n'(t) = (n-1)\lambda P_{n-1}(t) + (n+1)\mu P_{n+1}(t) - n(\lambda + \mu) P_n(t)$, with the initial condition $P_1(0) = 1$. The solution of these equations can be written as

$$P_0(t) = P_0 = \frac{\mu\, e^{(\lambda-\mu)t} - \mu}{\lambda\, e^{(\lambda-\mu)t} - \mu},$$

and

$$P_n(t) = (1 - P_0)\left(1 - \frac{\lambda}{\mu} P_0\right)\left(\frac{\lambda}{\mu} P_0\right)^{n-1}, \quad n = 1, 2, \ldots.$$

This distribution is a mixture of a point mass at zero and a geometric distribution modeling the number of trials until the first success where the probability of success is $p = 1 - \frac{\lambda}{\mu} P_0$.

(d) The mean of the process can be computed as

$$\mathbb{E}(X(t)) = (1 - P_0)\frac{1}{p} = \frac{1 - P_0}{1 - \frac{\lambda}{\mu} P_0} = e^{(\lambda-\mu)t}.$$

The variance is equal to

$$\mathbb{V}ar(X(t)) = (1 - P_0)\frac{1-p}{p^2} = \frac{\lambda + \mu}{\lambda - \mu} e^{(\lambda-\mu)t}\left(e^{(\lambda-\mu)t} - 1\right). \quad \square$$

7.2   Simulations in R

SIMULATION 7.1. Below we simulate a 20-step trajectory of a linear birth-and-death process with parameters $\lambda = 0.3$ and $\mu = 0.1$ that starts at time 0 in state 1. We generate two independent exponential times with rates $1/(n\lambda)$

and $1/(n\mu)$, and transition 1 unit up if a "birth" occurs before "death" or 1 unit down, otherwise.

```
#specifying parameters
lambda<- 0.3
mu<- 0.1
njumps<- 20

#setting state and time as vectors
N<- c()
time<- c()

#setting initial values
N[1]<- 1
time[1]<- 0

#specifying seed
set.seed(1022171)

#simulating trajectory
i<- 2

repeat {
time.birth<- (-1/(N[i-1]*lambda))*log(runif(1))
  time.death<- (-1/(N[i-1]*mu))*log(runif(1))

if(time.birth < time.death | N[i-1]==0) {
  time[i]<- time[i-1]+time.birth-0.001
    N[i]<- N[i-1]

  if(i==2*njumps+2) break
    else {
    time[i+1]<- time[i]+0.001
    N[i+1]<- N[i]+1
      i<- i+2
      }
    }
}
```

```
if(time.death < time.birth & N[i-1]!=0) {
  time[i]<- time[i-1]+time.death-0.001
    N[i]<- N[i-1]
  if(i==2*njumps+2) break
    else {
    time[i+1]<- time[i]+0.001
    N[i+1]<- N[i]-1
      i<- i+2
        }
 }


 }

#plotting trajectory
plot(time, N, type="l", lty=1, lwd=2, col=4, xlab="Time",
ylab="State", panel.first=grid())
```

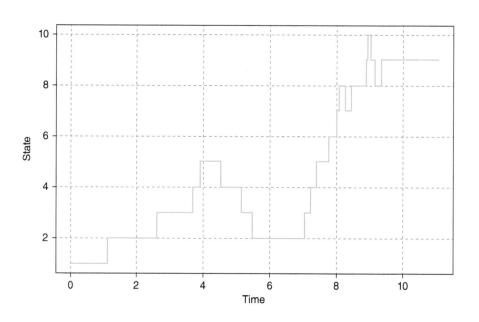

□

7.3 Applications of Birth-and-Death Process

APPLICATION 7.1. An *M/M/1 queue* is a birth-and-death process $\{X(t),\ t \geq 0\}$ with $\lambda_n = \lambda$ and $\mu_n = \mu$. We assume $\lambda < \mu$. In this process, customers join a queue (representing "births") at independent exponential times with mean $1/\lambda$, and leave the system (representing "deaths") after going through the service, which time is exponentially distributed with mean $1/\mu$. All customers act independently. In the name of the process, the first "M" means that the customer arrival process is Markovian, the second "M" refers to the fact that the service time is Markovian, and the "1" stands for a single server. An example of M/M/1 process is the number of customers in a bank with a single bank teller: customers enter the bank and join the line to the bank teller, then when it is their turn, they get a service from the teller and leave the bank. Another example is the number of broken cars in a repair shop with a single repairman: broken cars are added to the line to get a service from the repairman, and leave the shop once they are repaired.

(a) The Kolmogorov forward equations for M/M/1 process are

$$\begin{cases} P_0'(t) = -\lambda\, P_0(t) + \mu\, P_1(t), \\ P_n'(t) = \lambda\, P_{n-1}(t) + \mu\, P_{n+1}(t) - (\lambda + \mu)\, P_n(t),\ n = 1, 2, \ldots, \end{cases}$$

with the boundary condition $P_0(0) = 1$. The solution to these equations exists but is rarely used in practice. Instead, the *limiting* (or *steady-state*) probabilities are computed. They are defined as $\lim_{t\to\infty} P_n(t) = P_n$, $n \geq 0$. To find the limiting probabilities, we pass to the limit in the Kolmogorov forward equations as t tends to infinity. Replacing the left-hand side by 0 (since the derivative of a constant is 0), we obtain

$$\begin{cases} 0 = -\lambda\, P_0 + \mu\, P_1, \\ 0 = \lambda\, P_{n-1} + \mu\, P_{n+1} - (\lambda + \mu)\, P_n,\ n = 1, 2, \ldots. \end{cases}$$

These can be rewritten in the form of what is known as the *balance* (or *equilibrium*) *equations*:

$$\begin{cases} \mu\, P_1 = \lambda\, P_0, \\ (\lambda + \mu)\, P_n = \lambda\, P_{n-1} + \mu\, P_{n+1},\ n = 1, 2, \ldots. \end{cases}$$

The expression on the left represents the mean rate of leaving the state n, whereas the right-hand side gives the mean rate of entering state n. Thus, the balance equations equate the mean departure rate and the mean entrance rate.

To solve the balance equations, we notice that $\mu\, P_1 - \lambda\, P_0 = 0$ and $\mu\, P_{n+1} - \lambda\, P_n = \mu\, P_n - \lambda\, P_{n-1} = \ldots = \mu\, P_1 - \lambda\, P_0 = 0$. Therefore,

$$P_{n+1} = \frac{\lambda}{\mu} P_n = \left(\frac{\lambda}{\mu}\right)^2 P_{n-1} = \cdots = \left(\frac{\lambda}{\mu}\right)^{n+1} P_0, \quad n = 0, 1, \ldots \text{ Since all}$$

the probabilities must add up to 1, we conclude that $P_0 = \left[\sum_{n=0}^{\infty} \left(\frac{\lambda}{\mu}\right)^n\right]^{-1} =$

$1 - \frac{\lambda}{\mu}$, and so $P_n = \left(1 - \frac{\lambda}{\mu}\right)\left(\frac{\lambda}{\mu}\right)^n$, $n = 0, 1, 2, \ldots$. This is a geometric distribution modeling the number of failures before the first success, where the success probability is $p = 1 - \lambda/\mu$.

(b)   The average number of customers in the system, in the long run, is computed as

$$\lim_{t \to \infty} \mathbb{E}(X(t)) = \frac{1-p}{p} = \frac{1 - (1 - \lambda/\mu)}{1 - \lambda/\mu} = \frac{\lambda}{\mu - \lambda}.$$

(c) We will show that the amount of time T a customer spends in the system is an exponential random variable with mean $1/(\mu - \lambda)$. Indeed, suppose when the customer arrives, there are already n customers in the system. If $n = 0$, then T is the service time which is exponential with mean $1/\mu$. If $n > 0$, the customer will have to wait for one customer to complete the service, and then for n additional customers (including himself) to go through the service. Using the memoryless property of an exponential distribution, we conclude that the waiting time, in this case, is the sum of $n + 1$ independent exponential with mean $1/\mu$ random variables, which is a random variable having a gamma distribution with mean $(n + 1)/\mu$. We write

$$f_T(t) = \mathbb{E}\big[f_T(t \mid n \text{ customers})\big] = \sum_{n=0}^{\infty} \frac{t^n \mu^{n+1}}{\Gamma(n+1)} e^{-\mu t} P_n$$

$$= \sum_{n=0}^{\infty} \frac{(\mu t)^n}{n!} \mu e^{-\mu t} \left(1 - \frac{\lambda}{\mu}\right)\left(\frac{\lambda}{\mu}\right)^n = (\mu - \lambda) e^{-\mu t} \sum_{n=0}^{\infty} \frac{(\lambda t)^n}{n!} = (\mu - \lambda) e^{-(\mu - \lambda) t},$$

which is an exponential distribution with mean $1/(\mu - \lambda)$. From here, we can conclude that the probability that a customer spends more than time t in the system is $\mathbb{P}(T > t) = e^{-(\mu - \lambda)t}$.

(d) Consider an M/M/1 system in which arrivals occur with the rate $\lambda = 1$ per minute, and departures occur with the rate $\mu = 1.5$ per minute. The steady-state probabilities are

$$P_n = \left(1 - \frac{1}{1.5}\right)\left(\frac{1}{1.5}\right)^n = \left(\frac{1}{3}\right)\left(\frac{2}{3}\right)^n, \quad n = 0, 1, \ldots.$$

In the long run, there will be, on average, $\dfrac{\lambda}{\mu - \lambda} = \dfrac{1}{1.5 - 1} = 2$ customers in the system. The probability that in a long run, a customer will spend over 5 minutes in the system is $P(T > 5) = e^{-(1.5-1)(5)} = 0.082085.$ $\square$

Exercises

EXERCISE 7.1. A *Yule process*[2] (or a *linear birth process*) $\{X(t), t \geq 0\}$ is a birth-and-death process with $\lambda_n = n\lambda$ and $\mu_n = 0$, for all $n \geq 0$. In this process, each particle gives birth to one particle, independently of others, and never dies. It is a pure birth process.

(a) Suppose that initially the process is in state 1 (i.e., there is a single particle in the system). Show that the Kolmogorov forward equations (7.1) have the form $P_1'(t) = -\lambda P_1(t)$ and $P_n'(t) = (n-1)\lambda P_{n-1}(t) - n\lambda P_n(t)$, $n = 2, 3, \ldots$, with the boundary condition $P_1(0) = 1$.

(b) Verify that $P_n(t) = e^{-\lambda t}\left(1 - e^{-\lambda t}\right)^{n-1}$, $n = 1, 2, \ldots$. Note that it is a geometric distribution that models the number of trials until the first success where the probability of a success is $p = e^{-\lambda t}$.

(c) Show that the mean of the Yule process at time t is $\mathbb{E}(X(t)) = e^{\lambda t}$, and the variance is $\mathbb{V}ar(X(t)) = e^{\lambda t}(e^{\lambda t} - 1)$.

(d) If $\lambda = 4$ per week, what is the probability that there will be between 3 and 5 particles at week 1? What is the mean and standard deviation of the number of particles at week 1?

EXERCISE 7.2. Consider a Yule process, a birth-and-death process with parameters $\lambda_n = n\lambda$ and $\mu_n = 0$, for all $n \geq 0$. Suppose initially the process is in state m. That is, the initial size of the population is m particles.

(a) Verify that the Kolmogorov forward equations (7.1) have the form $P_m'(t) = -m\lambda P_m(t)$ and $P_n'(t) = (n-1)\lambda P_{n-1}(t) - n\lambda P_n(t)$, $n = m, m+1, \ldots$, with the boundary condition $P_m(0) = 1$.

(b) Verify that $P_n(t) = \binom{n-1}{n-m} e^{-m\lambda t}\left(1 - e^{-\lambda t}\right)^{n-m}$, $n = m, m+1, \ldots$. Note that it is a negative binomial distribution of the number of trials until the mth success, where the probability of a success is $p = e^{-\lambda t}$.

(c) Show that the mean of the Yule process at time t is $\mathbb{E}(X(t)) = m e^{\lambda t}$, and the variance is $\mathbb{V}ar(X(t)) = m e^{\lambda t}(e^{\lambda t} - 1)$.

(d) If there are originally 5 particles in the population and they multiply with rate $\lambda = 0.2$ per day, what is the probability that there will be exactly 12 particles on day 2? What are the average and standard deviation of the number of particles on day 2?

---

[2] Proposed in Yule, G. U. (1925). "A mathematical theory of evolution, based on the conclusions of Dr. J. C. Willis, F.R.S." *Philosophical transactions of the Royal Society of London. Series B, containing papers of a biological character*, 213: 21–87.

EXERCISE 7.3. A *linear death process* $\{X(t), t \geq 0\}$ is a birth-and-death process with the initial state N, and parameters $\lambda_n = 0$ and $\mu_n = n\mu$ for $n = 0, 1, \ldots, N - 1$.

(a) Show that the probabilities $P_n(t)$, $n = 0, 1, \ldots, N$, satisfy the Kolmogorov forward equations

$$\begin{cases} P_N'(t) = -N\mu\, P_N(t), \\ P_n'(t) = (n+1)\mu\, P_{n+1}(t) - n\mu\, P_n(t), \quad n = 0, 1, \ldots, N - 1. \end{cases}$$

(b) Verify that $P_n(t) = \binom{N}{n}\left(e^{-\mu t}\right)^n \left(1 - e^{-\mu t}\right)^{N-n}$, $n = 0, 1, \ldots, N$, which is a binomial distribution with parameters N and $p = e^{-\mu t}$.

(c) Show that the mean and variance of this process at time t are $\mathbb{E}(X(t)) = Ne^{-\mu t}$ and $\mathbb{V}ar(X(t)) = Ne^{-\mu t}(1 - e^{-\mu t})$.

(d) Assume $\mu = 0.02$ and $N = 15$. Find the probability that the process is in state 12 at time 3. What is the expected state at time 3? What is its standard deviation?

EXERCISE 7.4. Consider a linear birth-and-death process $\{X(t), t \geq 0\}$ described in Example 7.2.

(a) Suppose $\lambda = 1.3$ and $\mu = 0.2$. Compute the probability that the process will be in state 4 at time 2. Find the mean and variance of the process at time 2.

(b) Simulate a 50-step trajectory of this process, assuming $\lambda = 1.3$, $\mu = 0.2$, and the initial state is 1.

EXERCISE 7.5. Consider an M/M/1 queue described in Application 7.1.

(a) In the long run, how many customers do we expect to see in the system if $\lambda > \mu$? Explain on an intuitive level.

(b) Assume $\lambda = 3$ and $\mu = 5$. Find the probability that there will be more than 2 customers in the system in the long run.

(c) Compute the average number of customers in the system in the long run. Use $\lambda = 3$ and $\mu = 5$.

(d) Find the proportion of customers in the system in the long run who have to wait more than 1 minute.

EXERCISE 7.6. Suppose the bird count in a flock can be modeled as a birth-and-death process with *immigration* and *emigration*. In this model, $\lambda_n = n\lambda + \alpha$ and $\mu_n = n\mu + \beta$ where α is the rate of immigration (joining the flock), and β is the rate of emigration (leaving the flock). Generate a trajectory of the process until the flock size increases from 10 to 25 birds, assuming $\lambda = 1$, $\alpha = 0.3$, $\beta = 0.1$ and

(a) $\mu = 0.2$.
(b) $\mu = 0.8$.
(c) $\mu = 1$.
(d) $\mu = 1.2$.
(e) Discuss the difference in behaviors of the four trajectories. How many time units does each one span? Does the flock ever die out?

8

Branching Process

8.1 Definition of Branching Process

A discrete-time stochastic process $\{X_n,\, n \geq 0\}$ that gives the size of the nth generation of multiplying particles is called a *branching process* (or the *Bienaymé-Galton-Watson process*[1]). It starts with X_0 particles in the 0th generation. Each particle survives for one-time unit, at the end of which it splits into a random number of particles with a known probability distribution. The offspring particles survive for one-time unit, and produce a random number of offspring, independently from each other, and the process continues. Put formally, $X_n = \displaystyle\sum_{i=1}^{X_{n-1}} Z_i$ where Z_i is the size of the offspring of the ith particle in the $(n-1)$st generation. The distribution of the offspring size Z_i is identical for all particles, with $p_k = \mathbb{P}(Z_i = k)$, $k = 0, 1, \ldots,\ \mathbb{E}(Z_i) = \mu$, and $\mathbb{V}ar(Z_i) = \sigma^2$.

PROPOSITION 8.1. A branching process is a Markov chain.

PROOF: To prove that the Markov property holds, we use the fact that the distribution of the number of offspring for each individual particle is independent of the size of the current and all previous generations. We write

$$\mathbb{P}\big(X_n = j_n \,|\, X_0 = j_0,\, X_1 = j_1, \ldots, X_{n-1} = j_{n-1}\big)$$

$$= \mathbb{P}\Big(\sum_{i=1}^{j_{n-1}} Z_i = j_n \,|\, X_0 = j_0,\, X_1 = j_1, \ldots, X_{n-1} = j_{n-1}\Big)$$

[1] A French statistician Jules Bienaymé first addressed the problem of survival of family names in his 1845 article "De la loi de multiplication et de la durée des familles." *Société Philomatique de Paris. Extraits des Procès-Verbaux des Séances*: 37 – 39. This problem was rediscovered later by Englishman Sir Francis Galton who posed the problem (under the number 4001) in *The Educational Times and Journal of the College of Receptors*, Vol. XXV, No. 143, on page 300. The solution by English mathematician Reverend Henry William Watson was published in the same journal, Vol. XXVI, No. 148, on page 115.

$$= \mathbb{P}\Big(\sum_{i=1}^{j_{n-1}} Z_i = j_n \mid X_{n-1} = j_{n-1}\Big) = \mathbb{P}\big(X_n = j_n \mid X_{n-1} = j_{n-1}\big). \quad \square$$

PROPOSITION 8.2. Consider a branching process with a single initial ancestor, $X_0 = 1$.

(a) The mean of the size of the nth generation is $\mathbb{E}(X_n) = \mu^n$.

(b) The variance is $\mathbb{V}ar(X_0) = 0$, $\mathbb{V}ar(X_1) = \sigma^2$, and for $n \geq 2$,

$$\mathbb{V}ar(X_n) = \begin{cases} \sigma^2 \mu^{n-1}\left(\frac{1-\mu^n}{1-\mu}\right), & \text{if } \mu \neq 1 \\ \sigma^2 n, & \text{if } \mu = 1. \end{cases}$$

PROOF: (a) To find the expected value of X_n, we condition on the value of the size of the previous generation X_{n-1}. We write

$$\mathbb{E}(X_n) = \mathbb{E}\big[\mathbb{E}(X_n \mid X_{n-1})\big] = \mathbb{E}\Big[\mathbb{E}\Big(\sum_{i=1}^{X_{n-1}} Z_i \mid X_{n-1}\Big)\Big]$$

$$= \mathbb{E}\big(X_{n-1}\,\mathbb{E}(Z_i)\big) = \mu\mathbb{E}(X_{n-1}) = \mu^2\mathbb{E}(X_{n-2}) = \cdots = \mu^n\mathbb{E}(X_0) = \mu^n.$$

(b) The variance of $X_0 = 1$ is 0. The variance of $X_1 = Z_1$ is σ^2. To compute the expression for the variance of X_n for $n \geq 2$, we condition on the value of X_{n-1}. We obtain

$$\mathbb{V}ar(X_n) = \mathbb{V}ar\big[\mathbb{E}(X_n \mid X_{n-1})\big] + \mathbb{E}\big[\mathbb{V}ar(X_n \mid X_{n-1})\big]$$

$$= \mathbb{V}ar\Big[\mathbb{E}\Big(\sum_{i=1}^{X_{n-1}} Z_i \mid X_{n-1}\Big)\Big]$$

$$+ \mathbb{E}\Big[\mathbb{V}ar\Big(\sum_{i=1}^{X_{n-1}} Z_i \mid X_{n-1}\Big)\Big] = \mathbb{V}ar\big[X_{n-1}\mathbb{E}(Z_i)\big] + \mathbb{E}\big[X_{n-1}\mathbb{V}ar(Z_i)\big]$$

$$= \mathbb{V}ar(\mu X_{n-1}) + \mathbb{E}(\sigma^2 X_{n-1}) = \mu^2\,\mathbb{V}ar(X_{n-1}) + \sigma^2\mu^{n-1}$$

$$= \mu^2\big[\mu^2\,\mathbb{V}ar(X_{n-2}) + \sigma^2\mu^{n-2}\big] + \sigma^2\mu^{n-1} = \mu^4\,\mathbb{V}ar(X_{n-2}) + \sigma^2(\mu^{n-1}+\mu^n)$$

$$= \cdots = \mu^{2n-2}\,\mathbb{V}ar(X_1) + \sigma^2(\mu^{n-1} + \cdots + \mu^{2n-3}) = \sigma^2(\mu^{n-1} + \ldots + \mu^{2n-2})$$

$$= \sigma^2\mu^{n-1}(1 + \mu + \cdots + \mu^{n-1}) = \sigma^2\mu^{n-1}\Big(\frac{1-\mu^n}{1-\mu}\Big), \quad \text{if } \mu \neq 1,$$

and if $\mu = 1$, the variance is equal to $\sigma^2 n$. $\quad \square$

EXAMPLE 8.1. (a) Suppose that the distribution of the size of the offspring is *Bernoulli*$(\frac{1}{2})$. We know that $\mu = \frac{1}{2}$ and $\sigma^2 = \frac{1}{4}$. Therefore, the expected size of the nth generation is $\mathbb{E}(X_n) = \left(\frac{1}{2}\right)^n$ with the standard deviation

$$\sqrt{\mathbb{V}ar(X_n)} = \sqrt{\left(\frac{1}{4}\right)\left(\frac{1}{2}\right)^{n-1}\left(\frac{1-\left(\frac{1}{2}\right)^n}{1-\frac{1}{2}}\right)} = \sqrt{\left(\frac{1}{2}\right)^n\left(1 - \left(\frac{1}{2}\right)^n\right)}.$$

(b) Assume now that the distribution of the offspring size is $Binomial(2, 1/2)$. It means that $\mu = 1$ and $\sigma^2 = 1/2$. Hence, $\mathbb{E}(X_n) = 1$, and the standard deviation is $\sqrt{\mathbb{V}ar(X_n)} = \sqrt{\frac{n}{2}}$.

(c) Here we will assume that the distribution of Z_i's is $Binomial(3, 1/2)$. Thus, the mean $\mu = 3/2$ and variance is $\sigma^2 = 3/4$. The mean of the size of the nth generation, therefore, is found as $\mathbb{E}(X_n) = (3/2)^n$ and the standard

deviation is $\sqrt{\mathbb{V}ar(X_n)} = \sqrt{\left(\frac{3}{4}\right)\left(\frac{3}{2}\right)^{n-1}\left(\frac{1-\left(\frac{3}{2}\right)^n}{1-\frac{3}{2}}\right)} = \sqrt{\left(\frac{3}{2}\right)^n\left(\left(\frac{3}{2}\right)^n - 1\right)}.$ □.

Branching processes are classified according to the value of the mean size of the offspring. If $\mu < 1$, the process is called *subcritical*, if $\mu = 1$, it is called *critical*, and if $\mu > 1$, it is called *supercritical*.

The *probability of extinction* is defined as the probability that a branching process with one initial ancestor will eventually have no particles. Put mathematically, let π_0 denote the probability of extinction. Then

$$\pi_0 = \lim_{n\to\infty} \mathbb{P}(X_n = 0 \mid X_0 = 1).$$

PROPOSITION 8.3. (a) For a subcritical ($\mu < 1$) process, the probability of extinction is 1 ($\pi_0 = 1$). Intuitively, if not enough particles are being born, the population will surely go extinct.
(b) For a critical ($\mu = 1$) process, $\pi_0 = 1$ (so the population is guaranteed to become extinct), unless $Z_i = 1$. In this case, the population consists of a single particle throughout all generations.
(c) For a supercritical ($\mu > 1$) process, the extinction can happen with a positive probability, but this probability is less than 1. In fact, π_0 is the smallest positive solution of the equation

$$\pi_0 = \sum_{k=0}^{\infty} \mathbb{P}(extinction \mid X_1 = k)\, p_k = \sum_{k=0}^{\infty} \pi_0^k\, p_k.$$

Note that $\pi_0 = 1$ is always one of the roots, which helps reduce this equation by one degree in case it has a polynomial form. □

EXAMPLE 8.2. Going back to Example 8.1, we see that when the distribution of the offspring size is $Bernoulli(1/2)$ $(\mu = 1/2)$ or $Binomial(2, 1/2)$ $(\mu = 1)$, the population will go extinct with probability 1. However, in the case of $Binomial(3, 1/2)$, the extinction of the population is not a sure thing. It will happen with probability π_0 where π_0 is the smallest positive root of $\pi_0 = \pi_0^0 (1/2)^3 + \pi_0^1 (3)(1/2)^3 + \pi_0^2 (3)(1/2)^3 + \pi_0^3 (1/2)^3$, or, rewritten in the standard form, $\pi_0^3 + 3\pi_0^2 - 5\pi_0 + 1 = 0$. Since $\pi_0 = 1$ is a solution, the cubic equation is reduced to the quadratic one $\pi_0^2 + 4\pi_0 - 1 = 0$, which roots are $-2 - \sqrt{5}$ and $\sqrt{5} - 2$. The probability of extinction is the positive of the two roots, $\pi = \sqrt{5} - 2 = 0.236068$. □

EXAMPLE 8.3. Suppose a branching process $\{X_n, \ n \geq 0\}$ starts with a single particle and the particles multiply according to a geometric distribution with the probability mass function $p(x) = p(1 - p)^x$, $x = 0, 1, 2, ...$

(a) The mean of the geometric distribution is $\mu = \mathbb{E}(Z_n) = \frac{1-p}{p}$. The process is subcritical if $\mu < 1$, or $\frac{1-p}{p} < 1$, or $p > 0.5$. The process is critical if $\mu = 1$ or $p = 0.5$. The process is supercritical if $\mu > 1$, or $p < 0.5$.

(b) The average size of the nth generation is $\mathbb{E}(X_n) = \mu^n = \left(\frac{1-p}{p}\right)^n$. The variance of the geometric distribution is $\sigma^2 = \frac{1-p}{p^2}$, and hence, if $p \neq 0.5$,

$$\mathbb{V}ar(X_n) = \sigma^2 \mu^{n-1}\left(\frac{1 - \mu^n}{1 - \mu}\right) = \frac{1-p}{p^2}\left(\frac{1-p}{p}\right)^{n-1}\left(\frac{1 - \left(\frac{1-p}{p}\right)^n}{1 - \frac{1-p}{p}}\right)$$

$$= \frac{1}{2p - 1}\left(\frac{1-p}{p}\right)^n\left(1 - \left(\frac{1-p}{p}\right)^n\right).$$

If $p = 0.5$, $\mathbb{V}ar(X_n) = \sigma^2 n = \frac{(1-p)n}{p^2} = 2n.$

(c) To find the probability of extinction as a function of $p, 0 < p < 0.5$, we note that π_0 is the smallest positive solution of the equation

$$\pi_0 = \sum_{n=0}^{\infty} \pi_0^n p(1 - p)^n = p \sum_{n=0}^{\infty} (\pi_0(1 - p))^n = \frac{p}{1 - \pi_0(1 - p)}.$$

This is a quadratic equation $(\pi_0 - 1)(\pi_0 - \frac{p}{1-p}) = 0$. Therefore, $\pi_0 = \frac{p}{1 - p}$. For $p \geq 0.5$, $\pi_0 = 1$. □

8.2 Simulations in R

SIMULATION 8.1. Here we simulate generation sizes for subcritical, critical, and supercritical processes. We assume that each population starts with a single initial particle and the respective offspring distributions are binomial with parameters $n = 3$, $p = 0.2$ ($\mu = np = 0.6$), $n = 5$, $p = 0.2$ ($\mu = 1$), and $n = 3$, $p = 0.6$ ($\mu = 1.8$).

```
#subcritical branching process, mu=0.6
k<- 1
N<- c()
N[1]<- 1

set.seed(300168)
for (i in 2:100) {
N[i]<- sum(rbinom(N[i-1],3,0.2))
    if (N[i]==0) {
        break }
}

N
```

```
[1] 1 2 2 1 0
```

```
#critical branching process, mu=1
N<- c()
N[1]<- 1

set.seed(3554218)
for (i in 2:100) {
N[i]<- sum(rbinom(N[i-1],5,0.2))
    if (N[i]==0) {
        break }
}

N
```

```
[1]  1  3  6  6 10  6  7  2  3  3  2  1
[13] 1  0
```

```
#supercritical branching process, mu=1.8
N<- c()
N[1]<- 1
```

```
set.seed(965823)
for (i in 2:20)
N[i]<- sum(rbinom(N[i-1],3,0.6))


N

[1]     1     2     3     7    13    23
[7]    41    79   128   236   422   789
[13] 1435  2609  4764  8594 15565 27964
[19] 50574 91053
```

From the above simulation, we see that the subcritical process dies out quickly, the critical takes a bit longer to die out, but the supercritical process grows to a large number of particles and doesn't become extinct.   □

SIMULATION 8.2. In this simulation, we generate and plot a trajectory of a branching process, in which offspring distribution is binomial with parameters $n = 3$ and $p = 0.7$. We utilize a self-referencing function in R and request to plot five generations of the process. The code is as follows.

```
library(tidyverse)

#specifying parameters gen.max<- 5
prob<- 0.7

#specifying seed
set.seed(2443534)

#simulating trajectory
level.segment<- function(gen, y, branch.num) {
branch <- data.frame(x=c(), y=c(), xend=c(), yend=c())
gen.remaining<- gen.max-gen-1
if (gen.remaining < 0) return(branch)

if (branch.num > 0) {
branch<- rbind(branch, data.frame(x=gen, y=y, xend=gen+1,
yend=y), level.segment(gen=gen+1, y=y, branch.num=rbinom(1,
3, prob)))
}

if (branch.num > 1) {
branch<- rbind(branch, data.frame(x=gen, y=y, xend=gen+1,
yend=y+ 3^gen.remaining), level.segment(gen=gen+1,
y=y+3^gen.remaining, branch.num=rbinom(1, 3, prob)))
}
```

```
if (branch.num > 2) {
branch<- rbind(branch, data.frame(x=gen, y=y, xend=gen+1,
yend=y-3^gen.remaining), level.segment(gen=gen+1,
y=y-3^gen.remaining, branch.num=rbinom(1, 3, prob)))
}

branch
}

bp<- level.segment(0, 0, rbinom(1, 3, prob))

#plotting trajectory
plot(bp[,1], bp[,2], type="n", yaxt="n", xlim=c(0,5),
ylim=c(range(bp)), xlab="Generation", ylab="Branching
process", panel.first=grid())

segments(bp[,1], bp[,2], bp[,3], bp[,4], lwd=2, col="blue")
```

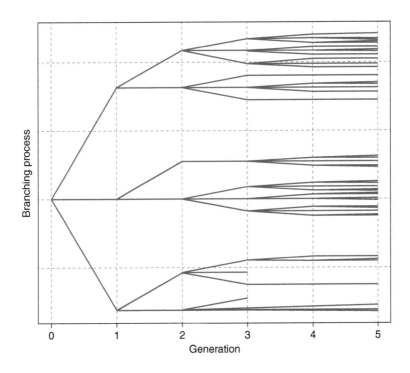

□

8.3   Applications of Branching Process

APPLICATION 8.1. One of the research interests of cultural anthropologists is that of a long-term history of population dynamics. Suppose a certain geographic area was initially populated by 25 families with 25 women of child-bearing age. In modern days, when a census of that area was taken, there were 19,856 women of child-bearing age. Ten percent of them had one daughter, 20% had two daughters, 60% had three daughters, and the others didn't have any daughters.

(a) From these data, we can estimate the mean size of female offspring. We have $\hat{\mu} = (0)(0.1) + (1)(0.1) + (2)(0.2) + (3)(0.6) = 2.3$. This is a supercritical process. The extinction of each family in this population is not going to happen for sure. In fact, the estimated probability of extinction of each family $\hat{\pi}_0$ solves $\hat{\pi}_0 = 0.1 + 0.1\,\hat{\pi}_0 + 0.2\,\hat{\pi}_0^2 + 0.6\,\hat{\pi}_0^3$, or, equivalently, $6\hat{\pi}_0^2 + 8\hat{\pi}_0 - 1 = 0$. The solution is $\hat{\pi}_0 = \frac{\sqrt{22}-4}{6} = 0.115069$. The estimated probability that all the 25 families become extinct is $\hat{\pi}_0^{25}$ which is very close to zero. However, the probability that at least one of the 25 families becomes extinct is computed as $1 - (1 - 0.115069)^{25} = 0.952931$.

(b) From the data, we can also assess how long ago the initial settlement took place. We know that on average, the size of the nth generation is $(25)(\hat{\mu}^n) = (25)(2.3)^n$. Therefore, $(25)(2.3)^{\hat{n}} = 19,856$. From here, $\hat{n} = \ln(19,856/25)/\ln(2.3) = 8.0169$ or about eight generations. Assuming that it takes around 25 years for a generation to mature, we can say that the settlement was established about $(8)(25) = 200$ years ago.   □

APPLICATION 8.2. In epidemiology, the most basic model of the spread of an infectious disease is a branching process. An initially infected individual will either recover, for instance, with probability 0.1 without infecting others, or will infect a zero-truncated Poisson(λ) random number of individuals where $\lambda = 2.4$, say.

(a) To find the mean of the nth generation of infected individuals, we note that the probability mass function for the offspring is $p(0) = 0.1$ and
$$p(n) = (0.9)\frac{\lambda^n}{n!}\frac{e^{-\lambda}}{1-e^{-\lambda}} = (0.9)\frac{(2.4)^n}{n!}\frac{e^{-2.4}}{1-e^{-2.4}}, \quad n = 1, 2, 3, \dots. \text{ The mean}$$
of this distribution is $\mu = (0.9)\dfrac{\lambda}{1-e^{-\lambda}} = \dfrac{(0.9)(2.4)}{1-e^{-2.4}} = 2.375501$. The

variance is $\sigma^2 = (0.9) \sum_{n=1}^{\infty} n^2 \dfrac{\lambda^n}{n!} \dfrac{e^{-\lambda}}{1 - e^{-\lambda}} - \mu^2 = \dfrac{0.9}{1 - e^{-\lambda}}\left(\lambda + \lambda^2\right) - \mu^2 =$

$\dfrac{0.9}{1 - e^{-2.4}}\left(2.4 + (2.4)^2\right) - (2.375501)^2 = 2.433697.$

(b) The expected number of infected individuals in the nth generation is $\mu^n = (2.375501)^n$, and the standard deviation is

$$\sqrt{\sigma^2 \mu^{n-1}\left(\frac{1 - \mu^n}{1 - \mu}\right)} = \sqrt{(2.433697)(2.375501)^{n-1}\left(\frac{1 - (2.375501)^n}{1 - 2.375501}\right)}$$

$$= \sqrt{(1.769317)(2.375501)^{n-1}\left((2.375501)^n - 1\right)}.$$

(c) Since $\mu > 1$, this is a supercritical process. The probability π_0 that the infection stops spreading is the smallest positive solution of the equation

$$\pi_0 = (0.1)(\pi_0)^0 + \sum_{n=1}^{\infty} \pi_0^n \, (0.9) \frac{(2.4)^n}{n!} \frac{e^{-2.4}}{1 - e^{-2.4}},$$

or

$$\pi_0 = 0.1 + \frac{0.9}{e^{2.4} - 1} \sum_{n=1}^{\infty} \frac{(2.4\pi_0)^n}{n!} = 0.1 + \frac{0.9}{e^{2.4} - 1}\left(e^{2.4\pi_0} - 1\right).$$

Solved numerically, $\pi_0 = 0.1340992$. The lines of code that produce this answer follow.

```
library(rootSolve)
equation<- function(x)
x-0.1-0.9/(exp(2.4)-1)*(exp(2.4*x)-1)

uniroot.all(equation, c(0,0.99))
```

0.1340992

□

Exercises

EXERCISE 8.1. Consider a colony of bacteria. Bacteria are known to reproduce asexually by binary fission, splitting into two identical cells. Suppose at time 0 the colony size is $X_0 = 100$ bacteria. Suppose also that at the end of a time unit, each bacterium, independently of the others, dies with probability 0.25, splits into two with probability 0.6, or continues living with probability 0.15.

(a) Show that the colony growth can be modeled as a supercritical branching process. Find the expected size of the nth generation and its variance.
(b) Compute the extinction probability for descendants of each bacterium.
(c) Find the probability that descendants of at least one of ten bacteria go extinct.

EXERCISE 8.2. Consider a branching process with a sole ancestor and a Poisson(λ) proliferation distribution.
(a) Determine the values of λ for which this process is supercritical (critical, subcritical).
(b) Give expressions for the mean and variance of the size of the nth generation.
(c) Give the equation that the extinction probability solves. Plot a graph of the numeric solution as a function of $\lambda > 1$.

EXERCISE 8.3. Based on work by Alfred J. Lotka[2] who analyzed the data from the 1920 U.S. Census, suppose that male offspring has a *zero-adjusted geometric distribution* of the form:

$$p(0) = 0.4828 \text{ and } p(n) = (0.228292)(0.5586)^{n-1} \text{ if } n = 1, 2, 3, ...$$

(a) Find the expected size of male offspring and its standard deviation.
(b) Consider a single male ancestor. Find the expected size of the nth generation of his descendants and its standard deviation.
(c) Compute the probability of extinction.

EXERCISE 8.4. Parlaying in gambling is defined as a series of bets in which winnings are used as a stake for further bets. This process can be modeled as a branching process. Suppose a gambler starts with a stake of $1, and can win $1 with probability 0.3, or $15 with probability 0.2, or $20 with probability 0.1, or $0 with probability 0.4.
(a) Show that this is a supercritical process.
(b) Find the expected winnings on the fifth bet.
(c) Find the probability that the gambler's stake eventually turns into $0.

EXERCISE 8.5. The spread of computer viruses is often modeled as a branching process. Assume that initially one computer is infected with a virus, and every day the virus is sent to other computers which number has a discrete uniform distribution between 0 and 3.

---

[2]Lotka, A. J. (1931a). "The extinction of families, I." *Journal of the Washington Academy of Sciences*, 21(16): 377 – 380; and Lotka, A. J. (1931b). "The extinction of families, II." *Journal of the Washington Academy of Sciences*, 21(18): 453 – 459.

(a) Is this process subcritical, critical, or supercritical?

(b) What are the average number and standard deviation of infected computers on day 10?

(c) Will the spread of viruses die out with probability 1? If not, find the extinction probability.

(d) Simulate the number of infected computers during 10 days. What is the simulated total number of infected computers?

EXERCISE 8.6. Consider a branching process that starts with a single particle in generation 1. Assume that the offspring has a distribution with the probability mass function $p(0) = 0.1$, $p(1) = 0.4$, $p(2) = 0.5$.

(a) Generate the size of the first 20 generations of this process. What is the size of the offspring of the 19th generation? How many total particles are in this population?

(b) Simulate and plot a sample trajectory of this process for the first six generations.

9

Brownian Motion

9.1 Definition of Brownian Motion

A stochastic process $\{B(t),\ t \geq 0\}$ is called a *standard Brownian motion*[1] (or a *Wiener process*[2]) if: (i) $B(0) = 0$, (ii) it has independent and stationary increments, and (iii) $B(t) \sim N(0, t)$, $t > 0$.

REMARK 9.1. Brownian motion is sometimes termed the *Bachelier process*. In his 1900 paper[3] a French mathematician Louis Bachelier derived the Brownian motion as the limit of random walks.

PROPOSITION 9.1. (a) Brownian motion is everywhere continuous but nowhere differentiable.
(b) Brownian motion hits every real number infinitely many times.

PROOF: The proofs of both statements are omitted. □

PROPOSITION 9.2. For a standard Brownian motion, the covariance between $B(s)$ and $B(t)$ is $\mathbb{C}ov(B(s), B(t)) = \min(s, t)$, $s, t \geq 0$.

---

[1] English botanist Robert Brown observed the movement of dust particles in liquid and described the motion in Brown, R. (1828). "A brief account of microscopical observations made in the months of June, July and August, 1827, on the particles contained in the pollen of plants; and on the general existence of active molecules in organic and inorganic bodies." *Philosophical Magazine, Series 2*, 4: 161 – 173.
[2] Norbert Wiener proposed a rigorous mathematical model for a Brownian motion in Wiener, N. (1921). "The average of an analytic functional and the Brownian movement." *Proceedings of the National Academy of Sciences of the United States of America*, 7(10): 294 – 298.
[3] Bachelier, L. (1900). "Théorie de la spéculation." *Annales Scientifiques de l'École Normale Supérieure*, 17: 21 – 86.

DOI: 10.1201/9781003244288-9

PROOF: Suppose $s \leq t$. Since the mean of a Brownian motion is zero and its increments are independent, we obtain

$$\mathbb{C}ov\big(B(s), B(t)\big) = \mathbb{E}\big(B(s)B(t)\big) = \mathbb{E}\big[B(s)\big(B(t) - B(s) + B(s)\big)\big]$$

$$= \mathbb{E}\big[B(s)\big(B(t) - B(s)\big)\big] + \mathbb{E}\big(B(s)\big)^2 = \mathbb{E}(B(s))\mathbb{E}\big(B(t) - B(s)\big) + \mathbb{E}\big(B(s)\big)^2$$

$$= \mathbb{E}\big(B(s)\big)^2 = \mathbb{V}ar\big(B(s)\big) = s = \min(s, t). \quad \square$$

PROPOSITION 9.3. (RESCALING RELATION). For any real a, $B(at)$ and $\sqrt{a}B(t)$ have the same distribution, where $B(t)$ denotes a standard Brownian motion.

PROOF: The distribution of $B(at)$ is $N(0, at)$ by the definition of a Brownian motion. Also, since $B(t) \sim N(0, t)$, we have that $\sqrt{a}B(t)$ is also normal with mean $\mathbb{E}(\sqrt{a}B(t)) = \sqrt{a}\,\mathbb{E}(B(t)) = 0$ and variance $\mathbb{V}ar(\sqrt{a}B(t)) = a\mathbb{V}ar(B(t)) = at$. $\quad \square$

EXAMPLE 9.1. Suppose we want to compute the conditional probability that a standard Brownian motion is below 3 at time 3, given that it is equal to 1 at time 1. We write

$$\mathbb{P}\big(B(3) < 3 \,|\, B(1) = 1\big) = \mathbb{P}\big(B(3) - B(1) < 3 - 1 \,|\, B(1) = 1\big)$$

$$= \mathbb{P}\big(B(3) - B(1) < 2\big) \;\; (by\;\; independence\;\; of\;\; increments)$$

$$= \mathbb{P}\big(B(2) < 2\big) \;\; (by\;\; stationarity\;\; of\;\; increments)$$

$$= \mathbb{P}\big(\sqrt{2}B(1) < 2\big) \;\; (by\;\; the\;\; rescaling\;\; relation)$$

$$= \mathbb{P}\big(B(1) < \sqrt{2}\big) = \Phi(\sqrt{2}) = 0.92135. \quad \square$$

Here and later $\Phi(\cdot)$ denotes the standard normal cumulative distribution function.

PROPOSITION 9.4. (DISTRIBUTION OF HITTING TIME). Denote by T_a the first time a standard Brownian motion hits a level $a > 0$. The cumulative distribution function of T_a is $F_{T_a}(t) = 2\big(1 - \Phi(a/\sqrt{t})\big)$, $t > 0$.

PROOF: We write $\mathbb{P}(B(t) \geq a) = \mathbb{P}(B(t) \geq a \,|\, T_a \leq t)\mathbb{P}(T_a \leq t) + \mathbb{P}(B(t) \geq a \,|\, T_a > t)\mathbb{P}(T_a > t)$. If $T_a > t$, $B(t)$ hasn't reached the level a yet, and it is impossible to have $B(t) \geq a$. Thus, the second term is equal to 0.

Also, from symmetry (see the picture), the Brownian motion is as likely to go up after hitting a as down, therefore,

$$\mathbb{P}(B(t) \geq a \mid T_a \leq t) = \mathbb{P}(B(t) \leq a \mid T_a \leq t) = \frac{1}{2}.$$

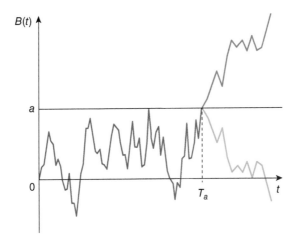

Consequently, $\mathbb{P}(B(t) \geq a) = \frac{1}{2}\mathbb{P}(T_a \leq t)$, and hence, $\mathbb{P}(T_a \leq t) = 2\mathbb{P}(B(t) \geq a) = 2(1 - \Phi(a/\sqrt{t}))$. $\square$

REMARK 9.2. In the proof above, we used the symmetry property of a Brownian motion. Put rigorously, this property is known as the *reflection principle* and is stated as: if a path of a Brownian motion reaches a value $B(s)$ at time s, a path after time s has the same distribution as its reflection about the value $B(s)$.

EXAMPLE 9.2. The probability that a Brownian motion reaches level 1 by time 5 is $\mathbb{P}(T_1 \leq 5) = 2(1 - \Phi(1/\sqrt{5})) = 0.6547.$ $\square$

PROPOSITION 9.5. (DISTRIBUTION OF MAXIMUM VALUE). Denote by $M(t)$ the maximum value that a Brownian motion attains on the interval $[0, t]$. The cumulative distribution function of $M(t)$ is $F_{M(t)}(a) = 2\Phi(a/\sqrt{t}) - 1$, for $a > 0$ and $t > 0$.

PROOF: Note that $M(t) > a > 0$, if and only if $T_a < t$. Therefore,

$$\mathbb{P}(M(t) \leq a) = 1 - \mathbb{P}(M(T) > a) = 1 - \mathbb{P}(T_a \leq t)$$
$$= 1 - 2(1 - \Phi(a/\sqrt{t})) = 2\Phi(a/\sqrt{t}) - 1. \quad \square$$

EXAMPLE 9.3. The probability that a Brownian motion is below 2 everywhere on the interval $[0, 4]$ is $\mathbb{P}(M(4) < 2) = 2\Phi(2/\sqrt{4}) - 1 = 2\Phi(1) - 1 = 0.682689.$ □

9.2   Processes Derived from Brownian Motion

9.2.1   Brownian Bridge

A *Brownian bridge* is a stochastic process $\{X(t), 0 \leq t \leq 1\}$ that satisfies the following properties: (i) $X(t)$ is normally distributed, (ii) $X(0) = X(1) = 0$, (iii) $\mathbb{E}(X(t)) = 0$, (iv) $\mathbb{V}ar(X(t)) = t(1 - t)$, and (v) $\mathbb{C}ov[X(s), X(t)] = \min(s, t) - st$, $0 \leq s, t \leq 1$.

We can think of a Brownian bridge as a Brownian motion on the interval $[0, 1]$, tied at the two ends. Note that the variance is equal to 0 at both ends of the interval as it should be since the values are deterministic at those points, and increases toward the middle, reaching its maximum at $t = 1/2$.

More generally, a Brownian bridge on the interval $[0, T]$, $\{X(t), 0 \leq t \leq T\}$, is such that: (i) $X(t)$ is normally distributed, (ii) $X(0) = X(T) = 0$, (iii) $\mathbb{E}(X(t)) = 0$, (iv) $\mathbb{V}ar(X(t)) = t(1 - t/T)$, and (v) $\mathbb{C}ov[X(s), X(t)] = \min(s, t) - st/T$, $0 \leq s, t \leq T$.

PROPOSITION 9.6. Suppose that $\{B(t), t \geq 0\}$ is a standard Brownian motion, and let $X(t) = B(t) - tB(1)$, $0 \leq t \leq 1$. Then $\{X(t), 0 \leq t \leq 1\}$ is a Brownian bridge.

PROOF: $X(t)$ has a normal distribution because $B(t)$ is normally distributed. The mean of $X(t)$ is $\mathbb{E}(X(t)) = \mathbb{E}(B(t)) - t\mathbb{E}(B(1)) = 0 - (t)(0) = 0$. Now, recall that $\mathbb{C}ov(B(s), B(t)) = \mathbb{E}(B(s)B(t)) = \min(s, t)$ (see Proposition 9.2). Assuming $s \leq t$, we compute the covariance between $X(s)$ and $X(t)$ as

$$\mathbb{C}ov(X(s), X(t)) = \mathbb{E}[B(s) - sB(1), B(t) - tB(1)]$$

$$= \mathbb{E}(B(s)B(t)) - t\mathbb{E}(B(s)B(1)) - s\mathbb{E}(B(1)B(t)) + st\mathbb{E}(B(1))^2$$

$$= \min(s, t) - t\min(s, 1) - s\min(1, t) + st\mathbb{V}ar(B(1))$$

$$= s - ts - st + st = s - st = \min(s, t) - st. □$$

REMARK 9.3. The proof of the above proposition can be extended (do it!) to show that $\{X(t) = B(t) - \frac{t}{T} B(T), t \geq 0\}$ is a Brownian bridge on the interval $[0, T]$.

9.2.2 Brownian Motion with Drift and Volatility

Let $\{B(t), t \geq 0\}$ denote a standard Brownian motion. A stochastic process $\{X(t) = \mu t + \sigma B(t), t \geq 0\}$ is called a Brownian motion with the *drift coefficient* μ and *volatility* (or *diffusion*) *coefficient* σ.

PROPOSITION 9.7. The distribution of $X(t)$ is normal with mean μt and variance $\sigma^2 t$. Also, the covariance between $X(s)$ and $X(t)$ is $\sigma^2 \min(s, t)$.

PROOF: The distribution of $X(t)$ is normal since $B(t)$ is normally distributed. The mean of $X(t)$ is $\mathbb{E}(X(t)) = \mu t + \sigma \mathbb{E}(B(t)) = \mu t$, and the variance is $Var(X(t)) = Var(\mu t + \sigma B(t)) = \sigma^2 Var(B(t)) = \sigma^2 t$. The covariance between $X(s)$ and $X(t)$ is $Cov(X(s), X(t)) = \mathbb{E}\Big((\mu s + \sigma B(s))(\mu t + \sigma B(t))\Big) - \mathbb{E}(\mu s + \sigma B(s))\mathbb{E}(\mu t + \sigma B(t)) = (\mu s)(\mu t) + \sigma^2 \mathbb{E}(B(s)B(t)) - (\mu s)(\mu t) = \sigma^2 \min(s, t)$. □

9.2.3 Geometric Brownian Motion

A stochastic process $\{Y(t) = Y(0) \exp(\mu t + \sigma B(t)), t \geq 0\}$ is called a *geometric* (or *exponential*) *Brownian motion*.

PROPOSITION 9.8. The distribution of $Y(t)$ is log-normal with the density function

$$f_{Y(t)}(y) = \frac{1}{\sqrt{2\pi\sigma^2 t}\, y} \exp\left(-\frac{(\ln y - \ln Y(0) - \mu t)^2}{2\sigma^2 t}\right), \quad y > 0.$$

The mean and variance are

$$\mathbb{E}(Y(t)) = Y(0)\, e^{\mu t + \sigma^2 t/2} \quad \text{and} \quad Var(Y(t)) = [Y(0)]^2 e^{2\mu t + \sigma^2 t}(e^{\sigma^2 t} - 1).$$

PROOF: Since $B(t) \sim N(0, t)$, the cumulative distribution function of $Y(t)$ is derived as follows:

$$F_{Y(t)}(y) = \mathbb{P}(Y(t) \leq y) = \mathbb{P}(\mu t + \sigma B(t) \leq \ln y - \ln Y(0))$$

$$= \mathbb{P}\left(B(t) \leq \frac{\ln y - \ln Y(0) - \mu t}{\sigma}\right) = \Phi\left(\frac{\ln y - \ln Y(0) - \mu t}{\sigma\sqrt{t}}\right), \quad y > 0.$$

The density is

$$f_{Y(t)}(y) = F'_{Y(t)}(y) = \frac{1}{\sqrt{2\pi\sigma^2 t}\, y} \, \exp\left(-\frac{(\ln y - \ln Y(0) - \mu t)^2}{2\sigma^2 t}\right), \quad y > 0.$$

Using the expression for the moment generating function of $B(t) \sim N(0, t)$, $\mathbb{E}(e^{\sigma B(t)}) = e^{\sigma^2 t/2}$, we get that the mean of $Y(t)$ is

$$\mathbb{E}(Y(t)) = \mathbb{E}(Y(0)\, e^{\mu t + \sigma B(t)}) = Y(0)\, e^{\mu t}\, \mathbb{E}(e^{\sigma B(t)}) = Y(0)\, e^{\mu t + \sigma^2 t/2},$$

and the variance is

$$\mathrm{Var}(Y(t)) = \mathbb{E}(Y(t))^2 - \left[\mathbb{E}(Y(t))\right]^2 = \mathbb{E}\left(Y(0)\, e^{\mu t + \sigma B(t)}\right)^2$$
$$- \left(Y(0)\, e^{\mu t + \sigma^2 t/2}\right)^2$$
$$= [Y(0)]^2\, e^{2\mu t}\, \mathbb{E}(e^{2\sigma B(t)}) - [Y(0)]^2\, e^{2\mu t + \sigma^2 t} = [Y(0)]^2\, e^{2\mu t + 2\sigma^2 t}$$
$$- [Y(0)]^2\, e^{2\mu t + \sigma^2 t}$$
$$= [Y(0)]^2\, e^{2\mu t + \sigma^2 t}\left(e^{\sigma^2 t} - 1\right). \quad \square$$

9.2.4   The Ornstein-Uhlenbeck Process

The Ornstein-Uhlenbeck process[4] $\{X(t),\ t \geq 0\}$ is a stochastic process of the form

$$X(t) = X(0)e^{-\theta t} + \mu\left(1 - e^{-\theta t}\right) + \frac{\sigma}{\sqrt{2\theta}}e^{-\theta t}B\left(e^{2\theta t} - 1\right).$$

Here μ is the drift, $\sigma > 0$ is the volatility, and $\theta > 0$ is an additional parameter.

PROPOSITION 9.9. The mean of $X(t)$ is $X(0)e^{-\theta t} + \mu\left(1 - e^{-\theta t}\right)$, and the variance is $\dfrac{\sigma^2}{2\theta}\left(1 - e^{-2\theta t}\right)$.

PROOF: The mean of $X(t)$ is

$$\mathbb{E}(X(t)) = X(0)e^{-\theta t} + \mu\left(1 - e^{-\theta t}\right) + \frac{\sigma}{\sqrt{2\theta}}e^{-\theta t}\, \mathbb{E}\left(B\left(e^{2\theta t} - 1\right)\right)$$
$$= X(0)e^{-\theta t} + \mu\left(1 - e^{-\theta t}\right),$$

---

[4]First appeared in Uhlenbeck, G. E. and L. S. Ornstein (1930). "On the theory of Brownian motion." *Physical Review*, 36: 823 – 841.

since the expected value of a Brownian motion (in this case, a time-transformed Brownian motion) is equal to 0. The variance of $X(t)$ is

$$\mathbb{V}ar(X(t)) = \frac{\sigma^2}{2\theta}e^{-2\theta t}\mathbb{V}ar\left(B\left(e^{2\theta t} - 1\right)\right) = \frac{\sigma^2}{2\theta}e^{-2\theta t}\left(e^{2\theta t} - 1\right)$$
$$= \frac{\sigma^2}{2\theta}\left(1 - e^{-2\theta t}\right).$$

□

REMARK 9.4. Note that as t increases, the mean of the Ornstein-Uhlenbeck process tends to μ. Therefore, the drift μ is the *long-term mean*, and the process is called *mean-reverting*. The parameter θ represents the rate by which the process reverts towards the mean. In addition, the variance of this process is bounded by a constant $\sigma^2/(2\theta)$, and in the long run, approaches this constant.   □

9.3   Simulations in R

SIMULATION 9.1. (ONE-DIMENSIONAL STANDARD BROWNIAN MOTION). The code below simulates three trajectories of a standard Brownian motion on the time interval that has 500 increments of size 0.01. The plot follows.

```
BM<- matrix(NA, nrow=500, ncol=3)

#specifying seed
set.seed(8221056)

#simulating trajectories
for (j in 1:3) {
BM[1,j]<- 0

for (i in 2:500)
  BM[i,j]<- BM[i-1,j] + sqrt(0.01)*rnorm(1)
}

#plotting trajectories
matplot(BM, type="l", lty=1, lwd=2, col=2:4,
ylim=c(range(BM)), xlab="Time", ylab="Brownian motion",
panel.first=grid())
```

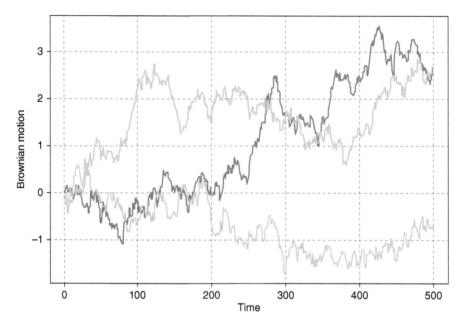

□

SIMULATION 9.2. (TWO-DIMENSIONAL BROWNIAN MOTION). A *two-dimensional Brownian motion* is a stochastic process that keeps track of two coordinates, both of which are independent Brownian motions. The R syntax below simulates and plots one trajectory of a two-dimensional Brownian motion.

```
BM<- matrix(NA, nrow=5000, ncol=2)

#specifying seed
set.seed(34885002)

#simulating two independent Brownian motions
for (j in 1:2) {
BM[1,j]<- 0

    for (i in 2:5000)
BM[i,j]<- BM[i-1,j] + sqrt(0.01)*rnorm(1)
}

#plotting trajectory
plot(x=BM[,1], y=BM[,2], type="l", col=4, xlab="x", ylab="y",
xlim=range(BM[,1]), ylim=range(BM[,2]), panel.first=grid())
```

```
#adding starting point
points(cbind(BM[1,1], BM[1,2]), pch=16, cex=2, col="green")

#adding ending point
points(cbind(BM[5000,1],BM[5000,2]), pch=16, cex=2,
col="red")
```

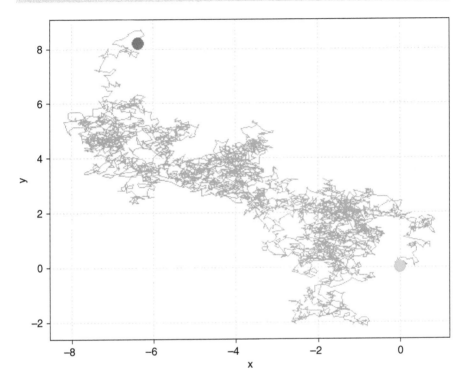

□

SIMULATION 9.3. (THREE-DIMENSIONAL BROWNIAN MOTION). A *three-dimensional Brownian motion* is a stochastic process that models position by three coordinates, defined by three independent Brownian motions. Below we simulate and plot a single trajectory of a three-dimensional Brownian motion.

```
nsteps<- 2000
BM<- matrix(NA, nrow=nsteps, ncol=3)

#specifying seed
set.seed(1133205)
```

```
#simulating three independent Brownian motions
for (j in 1:3) {
 BM[1,j]<- 0

   for (i in 2:nsteps)
BM[i,j]<- BM[i-1,j] + sqrt(0.01)*rnorm(1)
}

#plotting trajectory
library(plot3D)
lines3D(BM[,1], BM[,2], BM[,3], col=4, xlab="x", ylab="y",
zlab="z", xlim=range(BM[,1]), ylim=range(BM[,2]),
zlim=range(BM[,3]), bty="b2", ticktype="detailed")

#adding starting point
points3D(x=BM[1,1], y=BM[1,2], z=BM[1,3], add=TRUE, pch=16,
cex=2, col="green")

#adding ending point
points3D(BM[nsteps,1], BM[nsteps,2], BM[nsteps,3], add=TRUE,
pch=16, cex=2, col="red")
```

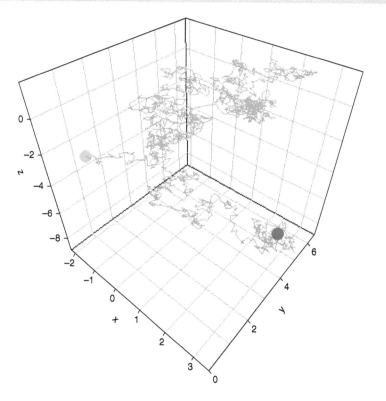

□

SIMULATION 9.4. (BROWNIAN BRIDGE). The following code generates three trajectories of a Brownian bridge. First we simulate three trajectories of a standard Brownian motion $\{B(t),\ t \in [0, 500]\}$, and then turn them into Brownian bridge trajectories by computing $\{X(t) = B(t) - \frac{t}{500}B(500),\ t \in [0, 500]\}$ (see Remark 9.3). The graphical output is given below.

```
#defining Brownian motion and Brownian bridge as matrices
BM<- matrix(NA, nrow=500, ncol=3)
BB<- matrix(NA, nrow=500, ncol=3)

#specifying seed
set.seed(76435567)

#simulating trajectories of Brownian motion
for (j in 1:3) {
BM[1,j]<- 0

for (i in 2:500)
BM[i,j]<- BM[i-1,j] + sqrt(0.01)*rnorm(1)
}

#computing trajectories of Brownian bridge
for(j in 1:3) {
for (i in 1:500)
BB[i,j]<- BM[i,j]-i/500*BM[500,j]
}

#plotting trajectories of Brownian bridge
matplot(BB, type="l", lty=1, lwd=2, col=2:4,
ylim=c(range(BB)), xlab="Time", ylab="Brownian bridge",
panel.first=grid())
```

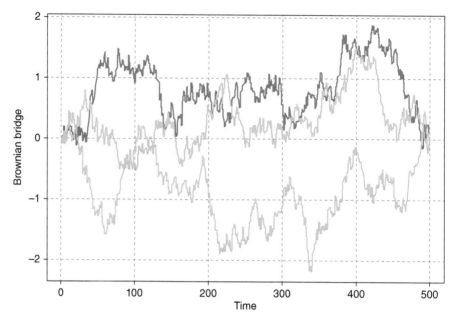

□

SIMULATION 9.5. (BROWNIAN MOTION WITH DRIFT AND VOLATIL-
ITY). Below we generate three trajectories of a Brownian motion with drift
$\mu = 1.3$ and volatility $\sigma = 0.5$. The code and plot follow.

```
#specifying parameters
mu<- 1.3
sigma<- 0.5

#defining Brownian motion as matrix
BM<- matrix(NA, nrow=500, ncol=3)

#specifying seed
set.seed(8463338)

#simulating trajectories
for (j in 1:3) {
BM[1,j]<- 0
```

```
for (i in 2:500)
BM[i,j]<- mu*0.01+BM[i-1,j] + sigma*sqrt(0.01)*rnorm(1)
}

#plotting trajectories
matplot(BM, type="l", lty=1, lwd=2, col=2:4,
ylim=c(range(BM)), xlab="Time", ylab="Brownian motion with
drift and volatility", panel.first=grid())
```

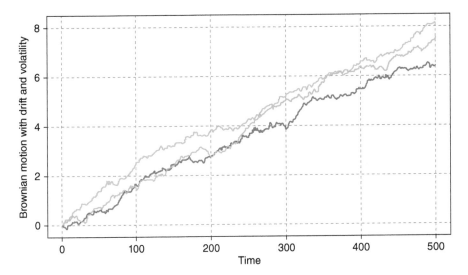

□

SIMULATION 9.6. (GEOMETRIC BROWNIAN MOTION). Here we simulate three trajectories of a geometric Brownian motion with the initial value $Y(0) = 2$, the drift coefficient $\mu = 1.3$, and volatility $\sigma = 0.5$. Since the trajectories of a Brownian motion with these values of drift and volatility have already been constructed in Simulation 9.5, all we need to do is to apply the exponential function to the simulated trajectories and multiply by the initial value. The code and graph follow.

```
#computing trajectories of geometric Brownian motion
GBMO<- 2
GBM<- GBMO*exp(BM)

#plotting trajectories
matplot(GBM, type="l", lty=1, lwd=2, col=2:4,
panel.first=grid(), ylim=c(range(GBM)), xlab="Time",
ylab="Geometric Brownian motion", panel.first=grid())
```

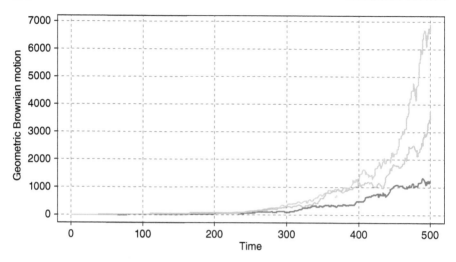

□

SIMULATION 9.7. (THE ORNSTEIN-UHLENBECK PROCESS). We base our simulation of a trajectory of an Ornstein-Uhlenbeck process $\{X(t),\, t \geq 0\}$ on the approximate recursive difference equation that it satisfies:

$$X(t + \Delta t) = X(t) + \theta\big(\mu - X(t)\big)\Delta t + \sigma\sqrt{\Delta t}B(1)$$

where $\Delta t > 0$ denotes a small increment of t. We omit the proof of this formula as it involves tedious algebra.

We use this relation with time increments $\Delta t = 1$ to simulate a trajectory for the values of the parameters $X(0) = 2, \theta = 0.8,\ \mu = 1.6$, and $\sigma = 0.5$. The code and graph follow.

```
#specifying parameters
theta<- 0.8
mu<- 1.6
sigma<- 0.5

#specifying seed
set.seed(2043442)

#defining Ornstein-Uhlenbeck trajectory as vector
OU<- c()

#specifying initial value
OU[1]<- 2
```

```
#simulating trajectory
for (i in 2:100)
OU[i]<- OU[i-1]+theta*(mu-OU[i-1])+sigma*rnorm(1)

#plotting trajectory
plot(1:100, OU, type="l", lty=1, lwd=2, col=4, xlab="Time",
ylab="Ornstein-Uhlenbeck process", first.panel=grid())
```

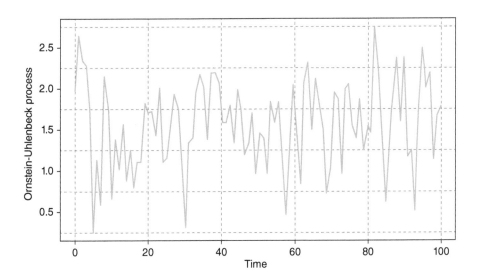

□

9.4 Applications of Brownian Motion

APPLICATION 9.1. Animal behavior researchers use the Brownian bridge to model movements of herds as they walk on their trails during daylight time and return to their designated lodging for an overnight stay. Suppose researchers observe the movements of a herd of deer during 8 hours of daylight. The main goal of the research is to estimate the distance between the north-most and south-most points that the deer have reached. This distance approximates the diameter of the deer home range. Assuming that the unit of measurement is one-tenth of a mile, below we compute the theoretical mean diameter of the home range in miles, and then simulate 1,000 trajectories to give an empirical estimate.

(a) It can be shown (see Exercise 9.9) that the expected value of the maximum of a Brownian bridge on the interval $[0, T]$ is $\frac{1}{2}\sqrt{\frac{\pi T}{2}}$. From symmetry, it can be argued that the minimum is expected to be of the same magnitude but with a negative sign. Therefore, the expected diameter of the deer home range is $\frac{1}{2}\sqrt{\frac{\pi T}{2}} - \left(-\frac{1}{2}\sqrt{\frac{\pi T}{2}}\right) = \sqrt{\frac{\pi T}{2}}$. We will assume that T is given in minutes, and thus, $T = (8)(60) = 480$ minutes. Hence, the theoretical value of the mean diameter of the home range is $\sqrt{\frac{\pi (480)}{2}} = 27.45873$ tenths of a mile or 2.75 miles.

(b) The code below simulates 1,000 trajectories of a Brownian bridge on the time interval $[0, 480]$ with an increment step of 1, and computes the sample mean of the range for the simulated trajectories.

```
#defining Brownian motion and Brownian bridge as matrices
BM<- matrix(NA, nrow=480, ncol=1000)
BB<- matrix(NA, nrow=480, ncol=1000)

#specifying seed set.seed(6769712)

#simulating trajectories of Brownian motion
for (j in 1:1000) {
BM[1,j]<- 0

  for (i in 2:480)
    BM[i,j]<- BM[i-1,j] + rnorm(1)
}

#computing trajectories of Brownian bridge
for(j in 1:1000){
  for (i in 1:480)
    BB[i,j]<- BM[i,j]-i/480*BM[480,j]
}

#computing ranges
range<- c()
for(j in 1:1000) {
range[j]<- max(BB[,j])-min(BB[,j])

#computing sample diameter of home range
print(diameter<- mean(range))
```

26.80793

Thus, the sample diameter of the home range is $26.80793/10 = 2.68$ miles.   □

APPLICATION 9.2. A geometric Brownian motion is often used to model the behavior of a stock price over time. The data set downloaded from *https://finance.yahoo.com/quote/AMZN/history/* contains Amazon.com, Inc. daily stock prices at the closing time of stock market exchange between 01/02/2020 and 06/30/2021, a total of 377 business days. First, we plot the data.

```
stock.data<- read.csv(file="./AMZN.csv", header=TRUE,
sep=",")

date<- as.POSIXct(stock.data$Date)
price<- stock.data$Close

#plotting stock price against date
plot(date, price, type="l", lwd=2, cex=0, col="light blue",
xlab="Time", ylab="Stock price", first.panel=grid())
```

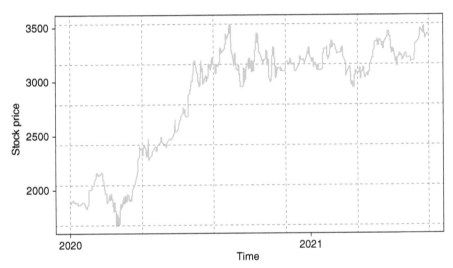

Now we estimate the parameters and simulate a trajectory of a geometric Brownian motion. The model for the stock price is

$$\big\{X(t) = X(t_1)\exp(\mu\,t + \sigma\,B(t)),\ t_1 \le t \le t_{377}\big\}.$$

We express the increments of the natural logarithm of the process as

$$\ln X(t_i) - \ln X(t_{i-1}) = \ln\frac{X(t_i)}{X(t_{i-1})} = \mu\,(t_i - t_{i-1}) + \sigma\,\big(B(t_i) - B(t_{i-1})\big).$$

We take the time increments of unit length, $t_i - t_{i-1} = 1$, and argue using the stationarity of increments that the log-ratios (or log-price increments) are distributed as $\mu + \sigma B(1)$. That is, they have a normal distribution with mean μ and variance σ^2. The code given below plots a histogram, estimates μ and σ by the sample values, simulates trajectories of the geometric Brownian motion, and plots the actual and simulated prices on the same graph.

```
#calculating increments of log-price
log.inc<- c()

price1<- price[-1]
price1.lag<- head(price, -1)
log.ratio<- log(price1/price1.lag)

#plotting histogram
library(rcompanion)
plotNormalHistogram(log.ratio, xlab="Log-price increments",
col="light blue")
```

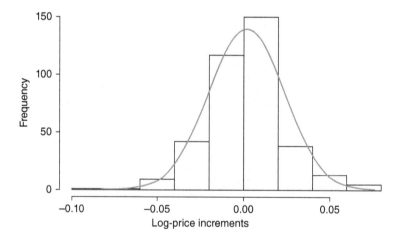

From the histogram, a bell-shaped curve reasonably describes the density of the log-price increments, thus we conclude that the distribution can be assumed normal.

```
#estimating parameters
print(mu.hat<- mean(log.ratio))
```

```
0.001581681
```

```
print(sigma.hat<- sd(log.ratio))
```

0.02156408

```
#specifying Brownian motion as vector
BM<- c()

#specifying initial value
BM[1]<- 0

#specifying seed
set.seed(43567347)

#simulating Brownian motion with drift and volatility
for (i in 2:377)
BM[i]<- mu.hat+BM[i-1] + sigma.hat*rnorm(1)

#computing values for geometric Brownian motion
GBM<-price[1]*exp(BM)

#plotting actual and simulated trajectories
plot(date, price, type="l", lty=1, lwd=2, col="blue",
xlab="Time", ylab="Stock price", first.panel=grid())
lines(date, GBM, lwd=2, col="green")
legend("bottomright", c("Actual price", "Simulated price"),
lty=1, col=c("blue", "green"))
```

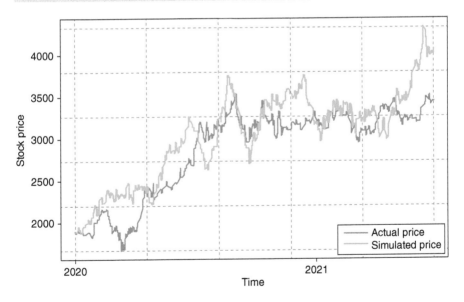

□

APPLICATION 9.3. A geometric Brownian motion has another very famous application in the financial world. In 1997, two American economists Myron Scholes and Robert Merton were awarded the Nobel Prize in Economics for the Black-Scholes-Merton Option Pricing model. In 1973, Fischer Black and Myron Scholes[5] published the derivation of the model and Robert Merton[6] expanded the results. Black died in 1995, so he wasn't awarded the Nobel Prize, for it is not given posthumously.

In this application, we discuss the model and derive the final formula. First, we introduce the key concepts.

In the financial market, a *stock option* is the right to buy (or sell) a stock at a predetermined price at a fixed time in the future.

An individual can buy (or sell) the stock at the price $X(s)$ at time $s < t$, and then sell (or buy) the stock at time t for the price $X(t)$. Suppose also that the individual can buy (or sell) at time 0 a stock *option* that gives him the right to buy a stock at time t for the price K per share. How much should he pay for one share of the stock option?

Suppose we loan out \$1 today with a risk-free interest compounded continuously at the fixed rate r. Then by time t, it would grow into amount \$$e^{rt}$. From here, we conclude that a \$1 at time t is worth \$$e^{-rt}$ in today's money. The rate r is termed the *discount factor*, and the function e^{-rt} is called the *discount function*. It represents the *present value* of an amount of \$1 at time t.

We will assume that it is a fair market and there is no opportunity for an *arbitrage*. That is, there is no opportunity for a sure profit. Under this assumption, the expected return of buying (selling) one share of stock at time $s < t$ and selling (buying) it at time t is zero, which translates into the identity involving the present values of the stock prices,

$$\mathbb{E}\left[e^{-rt} X(t) \mid X(u), 0 \le u \le s\right] = e^{-rs} X(s). \tag{9.1}$$

Turning to the stock option, if the price of one share of stock at time t is below K, it is not reasonable to excise the option. Therefore, the present value of the option is $e^{-rt}(X(t) - K)$, if $X(t) \ge K$, and 0, otherwise, which can be written as $e^{-rt}(X(t) - K)^+$.

---

[5]Black, F. and M. Scholes (1973). "The pricing of options and corporate liabilities." *Journal of Political Economy*, 81(3): 637 − 654.

[6]Merton, R. (1973). "Theory of rational option pricing." *Bell Journal of Economics and Management Science*, 4 (1): 141 − 183.

Let C be the price of one share of option at time zero. This is the quantity that we need to determine. In order not to create an arbitrage opportunity, we must have

$$\mathbb{E}\big[e^{-rt}(X(t)-K)^+ - C\big] = 0, \text{ or } C = e^{-rt}\mathbb{E}\big[(X(t)-K)^+\big]. \quad (9.2)$$

The Black-Scholes-Merton model assumes that $X(t)$ is a geometric Brownian motion $\{X(t) = X(0)\,e^{\mu t + \sigma B(t)},\ t \geq 0\}$ with the drift parameter μ and volatility σ. First, we see under what condition $\{X(t), t \geq 0\}$ satisfies (9.1), and then plug this process into (9.2) to derive the final expression for C.

Using the independence and stationarity of increments of the Brownian motion and utilizing the expression of its moment generating function, we write

$$\mathbb{E}\big[e^{-rt}X(t)\,|\,X(u),\, 0 \leq u \leq s\big] = e^{-rt}X(0)\,\mathbb{E}\big[e^{\mu t + \sigma B(t)}\,|\,B(u),\, 0 \leq u \leq s\big]$$

$$= e^{-rt}X(0)\,e^{\mu t + \sigma B(s)}\,\mathbb{E}\big[e^{\sigma(B(t)-B(s))}\big] = e^{-rt}X(s)\,e^{\mu(t-s)}\,\mathbb{E}\big(e^{\sigma B(t-s)}\big)$$

$$= e^{-rt}X(s)\,e^{\mu(t-s)+\frac{\sigma^2}{2}(t-s)} = e^{-rt+(\mu+\sigma^2/2)\,t-(\mu+\sigma^2/2)s}X(s) = e^{-rs}X(s),$$

if and only if $\mu + \sigma^2/2 = r$. This is the sought-for condition on the process $\{X(t),\ t \geq 0\}$. We now use it in the expression (9.2). We compute

$$C = e^{-rt}\mathbb{E}\big[(X(t)-K)^+\big] = e^{-rt}\int_{-\infty}^{\infty}\big(X(0)\,e^{\mu t+\sigma\sqrt{t}\,z}-K\big)^+\,\frac{1}{\sqrt{2\pi}}e^{-\frac{z^2}{2}}\,dz.$$

We want $X(0)\,e^{\mu t + \sigma\sqrt{t}\,z} - K \geq 0$, so $z \geq (\ln(K/X(0))-\mu t)/(\sigma\sqrt{t})$. Denote by $A = (\mu t - \ln(K/X(0)))/(\sigma\sqrt{t})$. Then the lower limit of integration is $-A$. We continue

$$C = X(0)\,e^{-(\mu+\sigma^2/2)t}\int_{-A}^{\infty}e^{\mu t+\sigma\sqrt{t}\,z}\,\frac{1}{\sqrt{2\pi}}e^{-\frac{z^2}{2}}\,dz$$

$$-\,e^{-rt}K\int_{-A}^{\infty}\frac{1}{\sqrt{2\pi}}e^{-\frac{z^2}{2}}\,dz$$

$$= X(0)\int_{-A}^{\infty}\frac{1}{\sqrt{2\pi}}e^{-\frac{(z-\sigma\sqrt{t})^2}{2}}\,dz - e^{-rt}K\int_{-A}^{\infty}\frac{1}{\sqrt{2\pi}}e^{-\frac{z^2}{2}}\,dz$$

$$= X(0)\int_{-\infty}^{A+\sigma\sqrt{t}}\frac{1}{\sqrt{2\pi}}e^{-\frac{z^2}{2}}\,dz - e^{-rt}K\int_{-\infty}^{A}\frac{1}{\sqrt{2\pi}}e^{-\frac{z^2}{2}}\,dz$$

$$= X(0)\,\Phi\big(A+\sigma\sqrt{t}\big) - e^{-rt}K\,\Phi(A).$$

To work with a numeric example, suppose the current price of one share of a stock is $X(0) = \$100$. Suppose the stock price can be modeled by the

Black-Scholes-Merton model with the drift coefficient $\mu = -0.45$ and volatility $\sigma = 1.1$. We want to compute the cost of the option to buy one share of the stock at time $t = 2$ for the cost of $K = \$120$. We write

$$A = \frac{\mu t - \ln(K/X(0))}{\sigma \sqrt{t}} = \frac{(-0.45)\,(2) - \ln(120/100)}{(1.1)\sqrt{2}} = -0.69574,$$

$$r = \mu + \frac{\sigma^2}{2} = -0.45 + \frac{(1.1)^2}{2} = 0.155,$$

and

$$C = X(0)\,\Phi\big(A + \sigma\sqrt{t}\big) - e^{-rt}\,K\,\Phi(A)$$
$$= (100)\Phi(-0.69574 + (1.1)\sqrt{2}) - e^{-(0.155)(2)}\,(120)\,\Phi(-0.69574) = \$59.09. \quad \square$$

APPLICATION 9.4. As opposed to stock and option prices that can rise indefinitely, interest rates and commodity prices move in a limited range. If their values are high, the demand drops, and consequently, the values drop. Likewise, if the values are low, demand increases, and eventually, the values increase. This characteristic is called a *reversion to a long-run mean*.

The Ornstein-Uhlenbeck (OU) process is a good mathematical model that captures this mean-reversion property. Below we fit the parameters of the OU process to a publicly available data set on daily natural gas prices between $1/4/2010$ and $8/11/2020$ (downloaded from *kaggle.com*). Recall that the OU process solves the difference equation

$$X(t + \Delta t) = X(t) + \theta(\mu - X(t))\Delta t + \sigma\sqrt{\Delta t}B(1).$$

Using $\Delta t = 1$, we can rewrite this equation as

$$X(t + 1) - X(t) = \theta\mu - \theta X(t) + \sigma B(1),$$

and note that this has the form of a linear regression of $X(t + 1) - X(t)$ on $X(t)$. Denoting by a and b the fitted intercept and slope, respectively, we can write $\widehat{\theta} = -b$ and $\widehat{\mu} = a/\widehat{\theta} = -a/b$. The volatility σ is estimated as the sample standard deviation of the error term. The code below estimates the parameters of the OU model and plots the observed and simulated trajectories.

```
gasprice.data<- read.csv(file="./gaspricedata.csv",
header=TRUE, sep=",")
```

```
#estimating parameters
inc<- gasprice.data$Price[-1]-head(gasprice.data$Price,-1)
fit<- glm(inc ~ head(gasprice.data$Price,-1))
theta.hat<- -fit$coefficients[2]
mu.hat<- fit$coefficients[1]/theta.hat
sigma.hat<- sigma(fit)

#specifying seed
set.seed(9467108)

#simulating OU process
OU<- c()
OU[1]<- gasprice.data$Price[1]

for (i in 2:length(gasprice.data$Date))
OU[i]<- OU[i-1]+theta.hat*(mu.hat-OU[i-1])+sigma.hat*rnorm(1)

#plotting trajectories
plot(as.Date(gasprice.data$Date), gasprice.data$Price,
type="l", lty=1, lwd=2, col=3, ylim=c(0,8), xlab="Time",
ylab="Natural gas price", first.panel=grid())
lines(as.Date(gasprice.data$Date), OU, lwd=2, col=4)
legend("bottomright", c("Actual price", "Simulated price"),
lty=1, col=3:4)
```

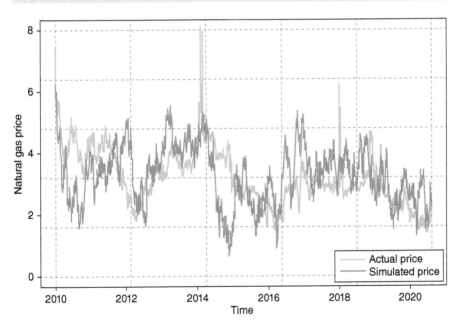

□

APPLICATION 9.5. Investors are interested in estimating correlation between various financial investments (for example, stock prices, stock option prices, bond yields, and commodity prices). To illustrate the concept of correlated Brownian motions, below we plot IBM stock prices at the closing of the stock market between 4/1/2020 and 3/30/2021, and U.S. 10-year treasury bond yields for the same time period. The data were downloaded from *https://www.investing.com*. The yields were rescaled by a multiplicative factor of 100 to plot comparable values.

```
data<- read.csv(file="./stock_bonds.csv", header=TRUE,
sep=",")

time<- as.Date(data$date)
IBM<- data$stock_price
bond<- data$bond_yield*100

#plotting the trajectories
plot(time, IBM, type="l", lty=1, lwd=2, col="blue",
ylim=c(0,200), xlab="Time", ylab="Stock price / Bond yield",
first.panel=grid())
lines(time, bond, lwd=2, col="orange")
legend("bottomright", c("IBM stock", "10-year bond"), lty=1,
lwd=3, col=c("blue", "orange"))
```

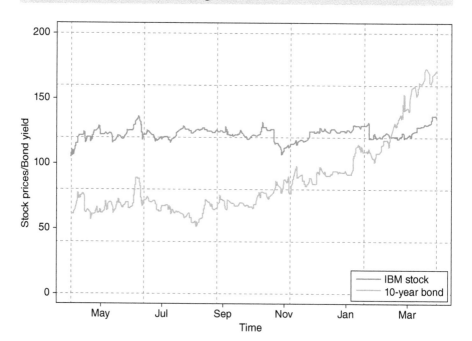

From the graph, the two curves exhibit somewhat similar behavior. To estimate the correlation coefficient between these two processes, we model them as correlated Brownian motions with the correlation coefficient ρ. To remove the time dependence, we resort to considering fixed-time increments where time steps are of size 1. The new processes are still correlated Brownian motions with the same correlation coefficient ρ (see Exercise 9.15 for proof). The sample Pearson correlation coefficient, computed on the increments, is the maximum-likelihood estimator of ρ. In our setting, the estimated correlation coefficient between IBM stock prices and 10-year treasury bond yields is about 0.36. The code and output follow.

```
#computing increments
IBM.diff<- IBM[-1]-head(IBM,-1)
bond.diff<- bond[-1]-head(bond,-1)

#estimating correlation coefficient
cor(IBM.diff, bond.diff)
```

0.3612484

☐

Exercises

EXERCISE 9.1. Let $\{B(t),\ t \geq 0\}$ be a standard Brownian motion. Show that the correlation between $B(s)$ and $B(t)$ is

$$\rho(B(s), B(t)) = \sqrt{\frac{\min(s,t)}{\max(s,t)}}, \quad s, t \geq 0.$$

EXERCISE 9.2. Show that the following processes are standard Brownian motions.
(a) $X(t) = tB(1/t)$ if $t > 0$, and 0 if $t = 0$, where $B(t)$ is a standard Brownian motion.
(b) $Y(t) = \alpha B_1(t) + \sqrt{1 - \alpha^2}\, B_2(t)$, $t \geq 0$, where $B_1(t)$ and $B_2(t)$ are independent standard Brownian motions, and $0 < \alpha < 1$.

EXERCISE 9.3. Let $\{B(t),\ t \geq 0\}$ be a standard Brownian motion. Find the probability that

(a) $0 < B(1) < 1$ and $1 < B(3) - B(1) < 3$.
(b) $0 < B(1) < 1$ and $1 < B(2) < 3$. Calculate numeric value in R.
(c) $0 < B(1) < 1$ and $0 < B(2) < \infty$.

EXERCISE 9.4. Let $\{B(t),\ t \geq 0\}$ be a standard Brownian motion. Suppose $0 \leq s < t$. Show that the distribution of $B(s) + B(t)$ is normal with mean 0 and variance $3s + t$.

EXERCISE 9.5. Let $\{B(t),\ t \geq 0\}$ denote a standard Brownian motion, and let $M(t) = \max_{0 \leq s \leq t} B(s)$. By Proposition 9.5, the cumulative distribution function of $M(t)$ is $F_{M(t)}(x) = 2\Phi(x/\sqrt{t}) - 1$, $x \geq 0$.
(a) Prove that $M(t)$ has the same distribution as $|B(t)|$, the absolute value of $B(t)$.

(b) Show that the expected value of $M(t)$ is $\mathbb{E}(M(t)) = \sqrt{\frac{2t}{\pi}}$.

(c) Compute the mean of the maximum on the interval $[0, 5]$. Simulate 1,000 trajectories of a standard Brownian motion on this interval and find an empirical estimate of the mean of the maximum.

EXERCISE 9.6. Let $\{B(t),\ t \geq 0\}$ denote a standard Brownian motion. Show that
(a) $\{X(t) = -B(t),\ t \geq 0\}$ is also a standard Brownian motion.
(b) $\mathbb{P}(\min_{0 \leq s \leq t} B(s) \leq x) = 2\Phi(x/\sqrt{t})$ where $x \leq 0$.
(c) Find the probability that the minimum of a standard Brownian motion is below -3 on the interval $[0, 5]$.
(d) Generate 1,000 trajectories of a standard Brownian motion on the interval $[0, 5]$ and find the sample probability that the minimum falls below -3.

EXERCISE 9.7. (a) Let $\{B(t),\ t \geq 0\}$ be a standard Brownian motion. Show that

$$X(t) = \begin{cases} (1-t)B\left(\frac{t}{1-t}\right), & \text{if } 0 \leq t < 1, \\ 0, & \text{if } t = 1 \end{cases}$$

is a Brownian bridge.

(b) Let $\{X(t),\ 0 \leq t \leq 1\}$ be a Brownian bridge. Show that

$$B(t) = (1+t)X\left(\frac{t}{1+t}\right),\ t \geq 0,$$

is a standard Brownian motion.

(c) Suppose $\{X(t), 0 \leq t \leq 1\}$ is a Brownian bridge and Z is a standard normal random variable independent of the Brownian bridge. Show that $B(t) = X(t) + t\,Z$ is a standard Brownian motion on $[0, 1]$.

EXERCISE 9.8. Let $\{B(t), t \geq 0\}$ be a standard Brownian motion.
(a) Show that for $0 \leq s < t$, the conditional distribution of $B(s)$, given $B(t)$, is normal with mean $\frac{s}{t} B(t)$ and variance $\frac{s}{t}(t - s)$.

(b) Let $B(t) = 0$. Argue that the process in part (a) is a Brownian bridge on the interval $[0, t]$. Note: This gives us another way to define a Brownian bridge, as a Brownian motion conditioned on the value at the endpoint.

EXERCISE 9.9. Let $\{B(t), 0 \leq t \leq T\}$ be a standard Brownian motion on the interval $[0, T]$, and let $M(T)$ denote the maximum of this process.
(a) Use the reflection principle to argue that $\mathbb{P}(M(T) \geq a, B(T) \leq x) = \mathbb{P}(B(T) \geq 2a - x), a > 0, x \leq a$.
(b) Show that the conditional density of $M(T)$ given that $B(T) = x$ has the form

$$f_{M(T)|B(T)}(a|x) = \frac{2(2a - x)}{T} e^{-\frac{2a(a-x)}{T}}, \ a > 0, \ x \leq a.$$

(c) Denote by $M_{BB}(T)$ the maximum of a Brownian bridge on the interval $[0, T]$. By the result of Exercise 9.8, the maximum of a Brownian bridge is the maximum of a Brownian motion conditioned on the value at time T. Use the formula derived in part (b) to show that the density of $M_{BB}(T)$ is

$$f_{M_{BB}(T)}(a) = \frac{4a}{T} e^{-\frac{2a^2}{T}}, \ a > 0.$$

(d) Show that the expected value of $M_{BB}(T)$ is $\frac{1}{2}\sqrt{\frac{\pi T}{2}}$.

EXERCISE 9.10. A herd of bison graze on a field and return to the water source once a day.
(a) Model the daily movement of the herd as a two-dimensional Brownian bridge, where both coordinates are independent Brownian bridges. Use minutes as time units. Plot a simulated trajectory.
(b) Suppose the home range of the herd is rectangular in shape and the linear distance unit is one-tenth of a mile. Find its expected area in square miles. Hint: Use the formula for the mean value of the diameter of a one-dimensional home range derived in Application 9.1.

(c) Simulate 1,000 trajectories of the daily movement of the herd and produce an empirical estimate of the area covered, assuming a rectangular shape of the home range. How does it compare to the theoretical value from part (b)?

EXERCISE 9.11. Ornithologists have collected data on bird population size in a bird viewing preserve for 5 years (60 months). The data are given in the table below.

Month	Size	Month	Size	Month	Size	Month	Size
1	10	16	232	31	472	46	888
2	14	17	276	32	510	47	927
3	47	18	331	33	510	48	898
4	50	19	346	34	523	49	949
5	60	20	348	35	594	50	979
6	91	21	369	36	565	51	994
7	118	22	405	37	634	52	1025
8	166	23	399	38	631	53	1003
9	119	24	410	39	671	54	962
10	123	25	460	40	744	55	968
11	109	26	400	41	782	56	991
12	160	27	432	42	786	57	982
13	168	28	458	43	773	58	959
14	216	29	460	44	784	59	973
15	240	30	478	45	842	60	974

These data can be modeled as a Brownian motion with drift and volatility.
(a) Plot the data.
(b) Compute the increments and construct a histogram. Are the increments normally distributed?
(c) Estimate the drift and volatility coefficients.
(d) Simulate a Brownian motion with the estimated parameters. Overlay the actual and simulated data on the same plot.

EXERCISE 9.12. The United States Environmental Protection Agency monitors an Air Quality Index (AQI) in a certain region. The data given below contain the values of AQI for 100 consecutive days for that region.

Day	AQI	Day	AQI	Day	AQI	Day	AQI	Day	AQI
1	108	21	48	41	40	61	31	81	24
2	98	22	43	42	35	62	28	82	22
3	86	23	43	43	31	63	25	83	22
4	78	24	39	44	30	64	24	84	20
5	85	25	49	45	26	65	31	85	23
6	77	26	46	46	24	66	29	86	21
7	70	27	49	47	27	67	22	87	20
8	63	28	44	48	25	68	20	88	27
9	58	29	46	49	33	69	28	89	35
10	53	30	42	50	32	70	25	90	32
11	44	31	44	51	39	71	23	91	36
12	40	32	40	52	35	72	22	92	33
13	47	33	48	53	35	73	28	93	25
14	48	34	43	54	32	74	25	94	23
15	46	35	43	55	27	75	21	95	20
16	50	36	39	56	20	76	14	96	18
17	47	37	40	57	24	77	22	97	13
18	42	38	36	58	31	78	27	98	12
19	38	39	43	59	29	79	34	99	19
20	46	40	47	60	26	80	31	100	17

(a) Plot the values of AQI against time (in days). Argue that the data may be modeled via a geometric Brownian motion.
(b) Estimate the parameters of the geometric Brownian motion model.
(c) Plot the actual and simulated values in the same coordinate system.

EXERCISE 9.13. The current price of a stock is $150. Suppose that the price of the stock changes according to a geometric Brownian motion with the drift coefficient $\mu = -0.4$ and variance $\sigma^2 = 0.76$. Use the Black-Scholes-Merton option pricing model to calculate the cost of an option to buy the stock at time $t = 7$ for a cost of $120.

EXERCISE 9.14. Foreign currency exchange rates can be modeled well with an Ornstein-Uhlenbeck (OU) process.
(a) Explain in simple words why the mean-reverting property and bounded variance are expected in this setting.
(b) The data file "Foreign_Exchange_Rates.csv" (*https://www.kaggle.com/brunotly/foreign-exchange-rates-per-dollar-20002019*) contains daily exchange rates for some currencies (as a ratio to US dollars) between 1/3/2000 and 12/31/2019. Select a currency and estimate the parameters of the OU process. Plot actual and simulated trajectories on the same graph.

EXERCISE 9.15. Let $\{B_1(t), t \geq 0\}$ and $\{B_2(t), t \geq 0\}$ be two independent standard Brownian motions. Consider a new process $\{B_3(t), t \geq 0\}$ formed as a linear combination of these two processes: $B_3(t) = \rho\, B_1(t) + \sqrt{1 - \rho^2}\, B_2(t)$ for some fixed ρ, $-1 < \rho < 1$.

(a) Show that $\{B_3(t), t \geq 0\}$ is a standard Brownian motion.

(b) Show that $\{B_1(t), t \geq 0\}$ and $\{B_3(t), t \geq 0\}$ are correlated with the correlation coefficient ρ.

(c) Show that for some fixed $s < t$, the increments $B_1(t) - B_1(s)$ and $B_3(t) - B_3(s)$ are correlated with the correlation coefficient ρ.

(d) Download historical data from *investing.com* website for financial investments of your choice, plot the two processes, and estimate the correlation coefficient between them.

Recommended Books

[1] Dobrow, R. P. *Introduction to Stochastic Processes with R*, Wiley, 2016.

[2] Ross, S. M. *Introduction to Probability Models*, Academic Press, 12th edition, 2019.

List of Notations

Index

Milton Keynes UK
Ingram Content Group UK Ltd.
UKHW022040141024
449569UK00014B/673

9 781032 154732